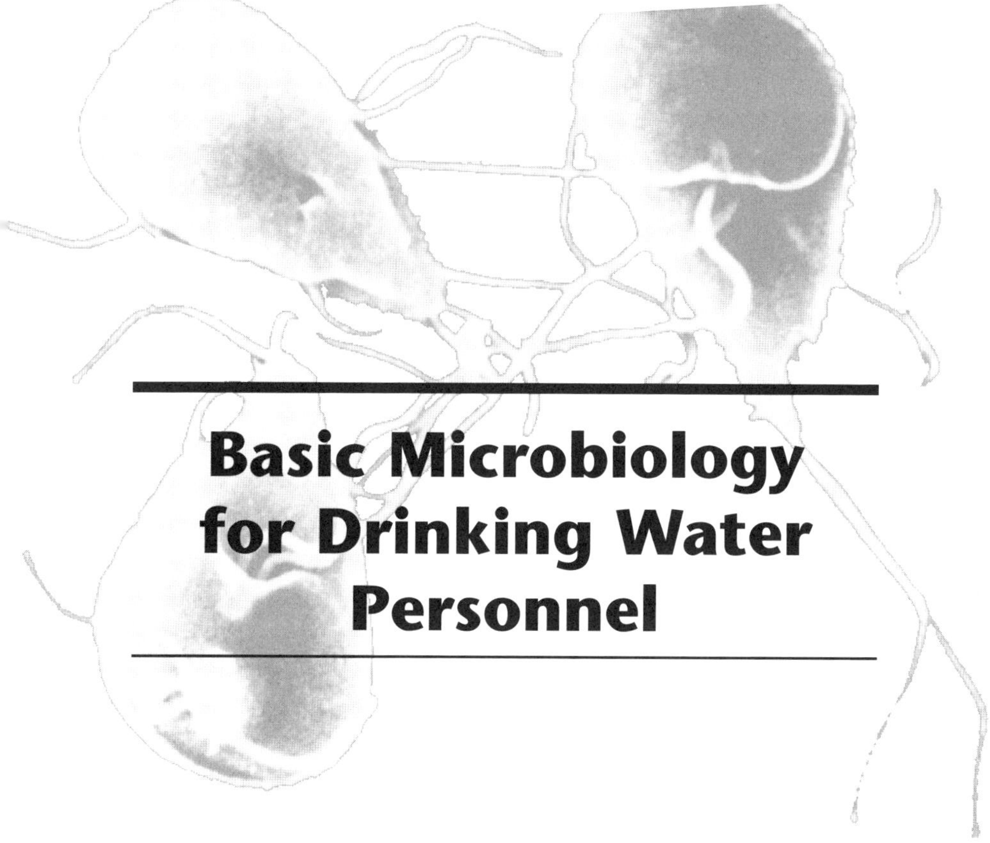

Basic Microbiology for Drinking Water Personnel

Basic Microbiology for Drinking Water Personnel

First Edition

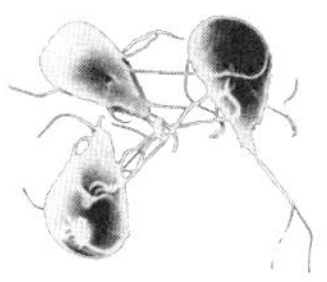

Dennis Hill

Des Moines Water Works

American Water Works Association

Dedicated to Safe Drinking Water

Basic Microbiology for Drinking Water Personnel

First Edition

Project manager: Melissa Christensen

Copy editors: Melissa Christensen and David Talley

Production editor: Carol Magin

Library of Congress Cataloging-in-Publication Data has been applied for.

Printed in the United States of America
American Water Works Association
6666 West Quincy Avenue
Denver, CO 80235

ISBN 1-58321-121-7

Printed on recycled paper

Table of Contents

Preface

My knowledge of microbiology and infectious diseases is a compilation of a microbiology degree from Iowa State University and twenty years of laboratory work in medical microbiology. Since 1994, I have been Des Moines Water Works' microbiologist.

Increased knowledge and general awareness of waterborne diseases by USEPA scientists and the public has made it important that water industry personnel become acquainted with basic microbiological concepts to better understand their jobs.

I chose to write a basic microbiology book to help those with little or no microbiology education understand the basic concepts of the science.

Des Moines Water Works' CEO and General Manager Dr. L.D. McMullen's interest in waterborne pathogens has made him especially supportive of laboratory science. This book resulted from his open support. I hope that it successfully serves the drinking water managers, operators, laboratory personnel, and others who read it.

Dennis Hill

1 Introduction

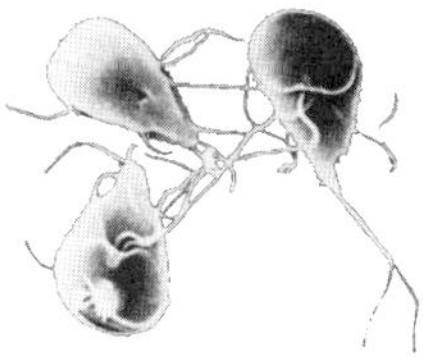

Like castles constantly under siege, human bodies are targets for continuous attacks by numerous infectious microorganisms. Humans have survived these assaults by developing biochemical and cellular defenses that make up the immune system.

We can prevent challenges to the immune system and stay healthy by avoiding harmful microorganisms. This practice requires careful personal hygiene and good nutrition, as well as social structures that preserve or clean the natural environment. Systems for drinking water treatment, sewage treatment, medical services, vaccination, and epidemiological intervention must complement wise ecological practices to limit the occurrences of disease epidemics.

Advances in medicine have nearly freed some parts of the world of many diseases that have plagued humanity until recent times. Still, smallpox, a once widespread and often fatal illness, is the only disease that has been totally eliminated through technological efforts. Smallpox was caused by a virus that existed exclusively in humans. A worldwide, concerted effort was needed to isolate and cure all people of the illness. After that, no animal or other environmental reservoir existed to allow reinfection. (Some countries have kept viable strains for research and germ warfare use.)

Unfortunately, human, animal, and environmental reservoirs still harbor many other infectious organisms that cause diseases like poliomyelitis, diphtheria, cholera, typhoid fever, and tuberculosis. These deadly threats still plague much of the world, and they could reemerge elsewhere if social, economic, or educational structures deteriorate or if political indifference compromises social, medical, or environmental systems designed to protect public health.

Disease Transmission

Disease organisms (called *pathogens*) vary in their ability to cause illness. Resistance to environmental factors varies among these pathogens. They also have different abilities to induce disease in a host, which is referred to as the organism's *virulence* or *pathogenicity*. Ingestion of only a few organisms can cause disease in some cases. Other organisms must be present in greater numbers before they can overwhelm a host's immune system. Sometimes tens of thousands of viruses or bacteria are required to induce a disease. Remember, however, that a needle hole can contain 500 million individual bacteria. A speck of *Salmonella* bacteria on a piece of meat or in a raw egg could make a person very ill, possibly with fatal consequences in cases involving children or the elderly. Thousands of different waterborne pathogens may enter a host in a single gulp of river or lake water.

Virulent microorganisms are often distinguished as either true pathogens or opportunistic pathogens. Their capabilities to cause disease can sometimes vary, but this terminology helps clarify the manner in which diseases develop. *True pathogens* are organisms that nearly always cause disease in their hosts. *Opportunistic pathogens* are those that take advantage of a host weakened by other illnesses, injuries, malnutrition, or physical anomalies. In other words, these organisms are pathogenic only when they encounter an opportunity. Opportunistic pathogens are often part of the normal bacterial flora present in the body of the host.

Human skin and intestinal tracts are always colonized with bacteria. Mucous membranes (such as in the lining of the mouth) also harbor a variety of bacteria. In someone who suffers from a viral cold for a week or two, bacteria from the oral cavity may take advantage of weakened defenses to cause ear infections, bronchitis, or pneumonia. This phenomenon accounts for the apparent re-emergence of a cold that many people have experienced. In reality, the relapse is often not the virus returning, but an opportunistic pathogen starting a new invasion.

Intestinal disease organisms are often contracted when a person handles previously contaminated objects such as a doorknob, then touches his or her mouth (hand-to-mouth transmission). Sometimes infected people transmit the germs through food that they have prepared, by shaking hands, or by otherwise coming into physical contact with others. In these cases, intestinal organisms are transferred by contact with feces through a process called the fecal–oral route of transmission. Intestinal illness may also be spread by ingesting spoiled food or drinking contaminated water.

Respiratory diseases are spread by aerosols arising from coughs, sneezes, and speech. People also pick up pathogens when they touch contaminated objects such as doorknobs or sink faucet handles and then touch their eyes and noses.

Crowded environments with limited ventilation encourage rapid person-to-person disease transmission. Day-care centers are notorious for disease transmission, because of the close contact and poor hygiene of the children.

Purification of drinking water is the most successful means of preventing the spread of disease in a society. Effective water purification eliminates waterborne pathogens, and the primary transmission mode of diseases is shifted to the person-to-person contacts previously mentioned. However, if treatment facilities fail in their goal of purifying water, large epidemics may result with serious consequences. Impeccable operation of treatment plants must combine with persistent watershed management to promote public health in today's industrialized world.

Monitoring to detect all bacterial, viral, and protozoan species that cause waterborne disease would be a very time-consuming and expensive undertaking. For this reason, a certain group of bacteria has been selected by the water industry for monitoring to assess the overall fecal contamination of the water. These bacteria comprise the *total coliform group*, which includes *Escherichia coli* as one of its members. These organisms are found in large numbers in mammalian intestines and therefore their feces. Coliforms are primarily opportunistic pathogens that are safe to ingest in small numbers. However, when they are detected in a water sample, the water is assumed to be unsafe to drink because of the possible presence of pathogenic microorganisms.

A group called *fecal coliforms* is often spoken of rather than *E. coli*. Fecal coliforms are primarily *E. coli* strains that thrive at the body temperature of humans.

Bacteriophages are viruses that attack only bacteria. *Coliphages* are bacteriophages that attack coliform bacteria. Because coliphages are specifically linked to coliforms, their presence indirectly indicates potentially contaminated water. Also, because the coliphage organisms are closely related to human viruses, they are used as indicators for inactivation of viruses.

Conventional tissue cultures for human viruses are time consuming and difficult. Therefore, USEPA has chosen to suggest assaying water for coliphages to determine a utility's ability to inactivate viruses.

Immunology

The science of immunology studies the numerous ways with which people ward off disease. The continuous struggle between host and invading microbe encourages an ever-developing immune potential. This fight encourages microorganisms to adapt also, leading to the rapid development of new strains and diseases.

Pathogenic microbes may be contracted from contact with humans or animals or from the environment. The human body offers many defenses against disease resulting from this contact. A respiratory virus first encounters viscous mucous in the nasopharynx that physically traps the organisms. Tiny hairs called *cilia* move the mucous outward, away from the lungs, to be swallowed or eliminated as nasal secretions. Coughing also helps physically clear the lungs of microbes trapped in mucous. Similarly, pathogens in the stomach are eliminated from the body by vomiting and intestinal pathogens by diarrhea. These functions are defensive responses generated by the body, not conditions caused by the disease organism.

IgA is an antibody that is secreted into the mucous. This antibody may attach to some invading organisms trapped there and inactivate them.

If a virus successfully attaches to a cell in the nasal cavity, it injects its chromosomal material (DNA or RNA) into the host cell. The virus chromosome then inserts itself into the human cell's chromosome and induces the manufacturing of new viruses. This process rapidly proceeds until the cell ruptures and disperses several viruses onto neighboring cells, where the infection continues and escalates.

A cellular chemical called *interferon* is secreted to prevent the infection from proceeding unchecked. Interferon helps to strengthen neighboring cells from viral attack and therefore impedes the progress of the infection.

White blood cells called *neutrophils* also contribute to the immune system's function of containing infection. They travel through the bloodstream until they encounter infected tissues, where they engulf and destroy viruses and bacteria.

Other white blood cells called *lymphocytes* detect the invading microorganisms as foreign to the body. In response, they produce antibodies, which attach to the invaders, inactivating and destroying them. Antibody production is a complex event, usually requiring a minimum of 2 weeks before enough are produced to effectively fight off infections. For this reason, the body must rely primarily on its early defense mechanisms previously described. Antibodies help most by warding off the recurrence of disease. In such instances, the immune system is already sensitive to the invading pathogen, allowing a more rapid antibody response.

Immune systems are poorly developed in infants, and antibody production is minimal until 2 years of age. Some antibodies pass from a mother's blood across the placenta and into the fetal blood system giving newborn babies a basic defense against disease. Breast milk is also rich in antibodies, so a mother who nurses her child supplies enhanced protection against disease along with nutrition.

Microbial Environmental Diversity

Microorganisms are the quintessential diversifiers. A single cell may multiply into billions overnight and all those cell divisions create a strong likelihood of mutations. The billions of gene variations involved in these mutations in turn allow a great adaptive ability. Variations in genes increase the probability that resulting characteristics will allow organisms to survive changes in environmental conditions. Microbes, therefore, tolerate varying conditions and capitalize on new opportunities for nutrition and growth.

Recent studies of bacteria and protozoa have produced a revolutionary understanding that expands the idea of adaptive potential due to mutations. The traditional view of microbes held that each species possessed definable characteristics and performed predictably in any one environment, except for mutant strains. Research has revealed, however, any one species may contain inactive genes, as well as active ones. When the microbe's environment changes, the inactive genes may become active, resulting in new characteristics. In other words, species need not undergo genetic change through mutation to survive. Inactive genes give organisms a stored potential to adapt when needed.

This discovery might powerfully affect the water industry, for example, when source water quality is allowed to deteriorate as a result of poor watershed management. If the change triggers inactive genetic potential, common microbes previously considered harmless may produce new pathogenic effects. A simple example would be a microbe that begins generating a toxin only when exposed to excessive nutrients (nitrates, phosphates, sewage) newly introduced to its environment. Such an event happened on the eastern US coast. A one-celled alga called *Pfiesteria* is very common in fresh and marine water. It remained harmless until communities next to marine bays allowed fertilizer runoff and human waste to pollute the water. Soon fish were attacked and killed, and human swimmers developed large open lesions. Scientists showed that *Pfiesteria* had unexpectedly begun (in response to the pollution) producing a potent neural toxin.

Such surprises may emerge in populations of other protozoa and bacteria, possibly precipitating outbreaks of waterborne illnesses. Water utilities must monitor this potential and carefully preserve healthy conditions of their watersheds.

The diversity and adaptability of bacterial metabolism gives these microbes astounding abilities to persist in the environment. They thrive in all types of soil and widely varying conditions. Some even colonize the hot springs of Yellowstone National Park.

Their resistance to adverse conditions may be most evident in their ability to grow on and inside the bodies of human hosts. The human immune system and skin oils maintain a constant fight against bacterial invasion. Despite this actively hostile environment, some invading pathogens still succeed in infecting hosts. Water industry personnel must remember the persistence of microbes and maintain constant vigilance.

Plan of the Book

The early chapters of this book provide overviews of important details about the various categories of microorganisms. Chapter 2 addresses the infectious bacteria most likely to pose challenges to water treatment processes, as well as bacteria that contribute to corrosion problems in the distribution system. Chapter 3 briefly summarizes the important characteristics of several other types of microorganisms: viruses, protozoa, algae, and multicellular organisms, such as nematodes, rotifers, small crustaceans, and insect larvae. Chapter 4 introduces analytical techniques for isolating and detecting different types of organisms. Chapter 5 presents a basic summary of applicable principles of chemistry and their implications for disinfection strategies.

2 Bacteria and Related Genera

Average size = 1/1,000 millimeter in diameter

Bacteria colonize every corner of the environment where humans live: homes and businesses, the soil outside the water supply, even the human body itself. Tiny as they are, bacteria can pose enormous threats to public health if conditions allow them to thrive and multiply.

Anaerobic and Aerobic Bacteria

The life processes of bacteria vary greatly from species to species. Some called *aerobic bacteria,* require oxygen for their metabolism. Some can grow without oxygen, giving them the name *anaerobic bacteria.* Still others are able to flourish under both conditions because of their *facultative* metabolism.

Primitive bacteria developed with anaerobic metabolism that allowed them to grow without oxygen, which was lacking in the environment at the time. However, microbe and plant metabolism produced oxygen as an end product, and over many centuries the atmosphere became oxygen rich.

This change limited the ecological niches in which the anaerobes could grow and survive. They could not tolerate oxygen's highly oxidative effect, but they survived and remain common in low-oxygen soils, as well as in animal and human mouths and intestines.

Eventually, bacteria able to withstand contact with oxygen evolved. They developed the ability to metabolize the gas, allowing them to derive more energy from nutrients they absorbed. Many of the bacteria found in the environment and within the bodies of animals and humans fall into this facultative group.

Strictly aerobic bacteria have also developed. Their lack of anaerobic abilities restricts their growth to oxygen-rich environments. Despite this restriction, they represent some of the most widespread water and soil organisms.

Bacteria are simple, unicellular organisms. (Among life forms, lonely viruses are smaller and less complex.) Most are free-living organisms, but a few require animal or plant hosts for survival. Bacteria absorb nutrients from their environments, excrete waste products, and secrete various toxins that help them invade tissues. Bacteria have no enclosed nucleus. Their chromosomal material is in the form of a large loop, packed into the cytoplasm of the cell.

The most common shapes of bacteria include rod, cocci (round), and spiral forms. Cellular arrangements occur singularly, in chains, and in clusters. Some species have one to numerous hair-like projections called *flagella* that allow the bacteria to swim, making them *motile* organisms.

Reactions to a special stain called the *Gram stain* separate two major groups of bacteria, Gram-positive and Gram-negative organisms. The former have thick walls, made primarily of peptidoglycan and stain purple. Gram-negative bacteria have thinner walls and stain pink. A microbiologist might characterize a bacteria sample as a Gram-negative rod, a Gram-positive cocci, and so forth.

Toxic Effects

Bacteria absorb nutrients from their environments and excrete waste products. They produce a great variety of enzymes that allow them to metabolize the nutrients, and some of these chemicals also allow them to break down and invade the tissues of plants and animals. Because bacteria damage the tissues of their hosts, they are considered pathogens.

Toxic enzymes secreted by a living bacterium are called *exotoxins*. Exotoxins, most of them products of Gram-positive bacteria, represent some of the most powerful toxins found in nature. Some disrupt connective tissues, while others impair neural and muscular activity or protein synthesis. Examples of these activities are seen in gas gangrene, tetanus (lockjaw), and botulism. Many other less severe infective processes also result from bacterial exotoxin production by bacteria, such as inflamation of a cut on a finger.

Another class of toxins, *endotoxins*, results only from Gram-negative bacteria. These substances are components of the bacteria cell walls, so they are not released until the bacterium dies. As it disintegrates, the harmful chemicals are dispersed into the host's tissues. Endotoxin activity interferes with various metabolic processes of the host's body, possibly leading to serious consequences such as blood clotting and lowered blood pressure, leading to shock and death.

Transmission of bacterial diseases can be from person-to-person, ingestion of contaminated food or drink, or from physical contact with animals and their waste products. Aggressively pathogenic bacteria ("true pathogens") are transmitted in these manners.

Because bacteria can live on and in the human body as "normal flora," the phenomenon of opportunistic pathogenicity is also common. This happens when harmless bacterial strains become pathogenic if the host's body is weakened by an injury, another infectious pathogen, or physiological disease. Generally weakened health, such as with the elderly, also helps aid opportunism.

People today seldom realize how they benefit from antibiotics. Without a dab of antibiotic salve applied to an infected cut, or without antibiotic pills taken for bronchitis, urinary tract infection, or an abscessed tooth, etc., many of us would suffer greatly or die. Our knowledge of disease transmission along with good hygiene, good environmental practices, and proper drinking water and wastewater treatment is also paramount in protecting us from bacterial diseases.

Some bacterial pathogens are common in raw water sources; however, if purification systems remain functional, traditional drinking water treatment methods will effectively kill the pathogens.

Escherichia coli

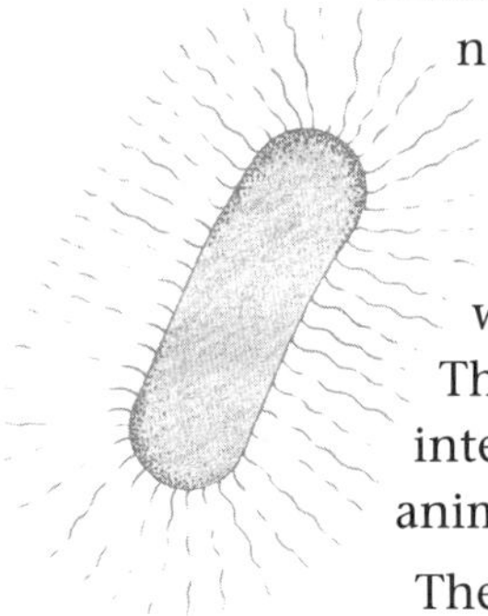

E. coli is a Gram-negative rod of the family Enterobacteriaceae. Most strains of *E. coli* do not produce debilitating toxins; however, clinically, they are very common, opportunistic pathogens, especially when encountered in abdominal wounds and urinary tract infections. They are found in high numbers in the intestines of humans and warm-blooded animals.

There are a few *E. coli* strains that are aggressively pathogenic or toxigenic. Enteropathogenic *E. coli* causes gastroenteritis, and is usually contracted via contaminated food. *E. coli* O157:H7 is a toxigenic strain that first causes gastroenteritis, and eventually may produce sufficient toxin to damage its host's kidneys, a condition known as hemolytic uremic syndrome. Children are especially susceptible to this strain, and may die despite intensely applied medical treatment.

E. coli O157:H7 is found in cattle feces and consequently may contaminate processed meats. Ground hamburger is most notorious and should be thoroughly cooked to avoid contraction of this *E. coli* strain.

Contaminated water is also a possible source of this organism. Cattle yards with herds that are infected with the strain may have their wastes washed into source water streams by heavy rains. This would make the bacteria available for intake into a water facility.

Shigella

Shigella species are true pathogens. They are Gram-negative rods of the Enterobacteriaceae family. All *Shigella* species cause bacillary dysentery (shigellosis), an aggressive type of bloody diarrhea.

Shigella dysenteriae is especially pathogenic and can cause disease with a fatality rate of up to 20 percent. It is uncommon in the United States, though it flourishes in underdeveloped countries where flies, poor hygiene, and exposed food in open markets aid in its transmission.

Salmonella

There are numerous species and strains of *Salmonella*. They are Gram-negative rods belonging to the Enterobacteriaceae family. They can be found in humans and most animal species.

Salmonella bacteria are common inhabitants of chicken and turkey intestines. When the foul lay eggs, the shell becomes contaminated with their feces. Though an intact egg is quite impermeable to bacterial invasion, some eggs have minute cracks that allow entry by the *Salmonella*. Once inside, the bacteria have a very nutritious environment in which to multiply. If the contaminated egg is used for cooking, it will be a potential source of food poisoning unless the food is thoroughly cooked.

If chicken or turkey meat is not cooked well, or if the plates and utensils that touched the raw meat are used on the cooked meat, contraction of *Salmonella* becomes possible.

Unpasteurized milk products may also harbor *Salmonella*.

Rivers and lakes may contain *Salmonella* because of the many animals that carry the bacteria. Manure runoff from chicken and turkey businesses are potential point sources of water contamination.

The symptoms of *Salmonella* food poisoning—nausea, vomiting, and severe diarrhea—dramatically occur six to eight hours (or longer with some strains) after eating the contaminated food. Children, the elderly, and people who are otherwise weakened in their health may suffer severely and possibly die.

The *Salmonella*, which grows in its host's intestines, may also cause septicemia. Septicemia is a dangerous invasion of the bloodstream by the organism. This condition calls for the most intensive treatment and is the syndrome from which most people die.

Once the disease is over, some people and animals continue to harbor this organism in their intestines, without symptoms. The organism may find its way into the person's gallbladder and induce what is called a carrier state. *Salmonella* carriers risk contaminating food if they work in cafeterias or food-manufacturing plants.

Salmonella typhi causes typhoid fever. This species is rare in the United States and resides in infected humans only. This is fortunate, because it helps limit the occurrence of this potentially fatal infection. *S. typhi* has a greater tendency to cause severe septicemia than do the other *Salmonella* species. Contaminated food and water supplies transmit the organism.

In the early 1900s a cook, who eventually was given the name of Typhoid Mary, became an identified carrier of *S. typhi*. She moved from job to job under false identities. She transmitted the disease to 51 people, causing three deaths over a period of eight years. Eventually she was kept institutionalized until her death 25 years later.

Today, antibiotic treatment and the surgical removal of gallbladders cure patients of their carrier state.

Aeromonas and *Plesiomonas*

The genera of *Aeromonas* and *Plesiomonas* are Gram-negative rods. They primarily reside in aquatic environments worldwide, though *Plesiomonas* prefer warmer climates.

Both genera cause human intestinal illness if consumed and must be purified from drinking water.

Vibrio cholerae

V. cholerae is a curved, Gram-negative rod that is a true pathogen. It is not a normal part of the human intestinal flora, but rare individuals may carry it without showing symptoms of the disease. This carrier state creates a potential for disease continuation in humans. The individual carrying the bacteria may at a future time contaminate a body of water from which others may drink. *V. cholerae* is also able to live in brackish water and marine water.

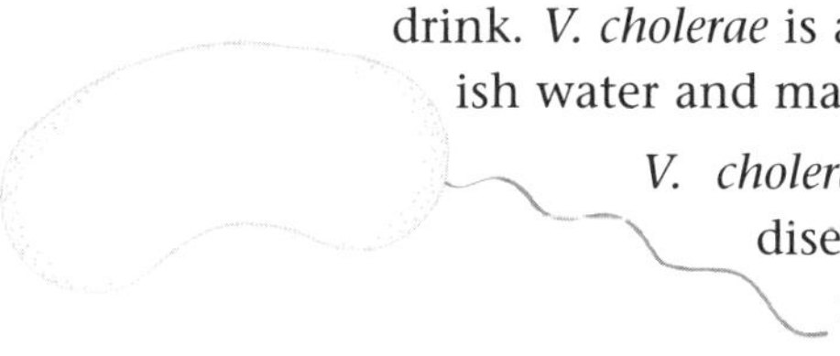

V. cholerae causes the diarrheal disease cholera. Cholera is a mild disease for many people that contract it; however, for some, it becomes severe and ends fatally if medical treatment is not administered. It causes epidemics and pandemics, especially in countries where sanitation is poor and sociopolitical stress occurs. This disease frequents war and refugee camps, compounding the misfortune of the people.

V. cholerae releases several toxins into the intestines. Some cause extensive secretion of fluids and electrolytes (sodium and potassium) from the intestinal walls. This causes the patient to dehydrate rapidly, and potentially experience fatal shock. Young children are especially susceptible.

When epidemics occur, the organism is transmitted via the fecal–oral route. Contaminated water is also a source, especially if it is drunk untreated.

Cholera is common in Africa and Asia. Recently, there have also been epidemics of cholera in Central and South American countries, where it had been absent for over a century. In desperation, the people affected occasionally suspect political opponents for their disease, but undesirable social and environmental conditions are the actual precipitators of the epidemics.

Campylobacter

Campylobacter is a genus of curved Gram-negative rods that is not found as normal intestinal flora in humans. It is commonly carried by various animals, including poultry, dogs, cats, sheep, and cattle.

Campylobacter species most commonly cause gastroenteritis. They can also cause dental disease, and systemic infections of the brain, heart, and joints accompanied by fever.

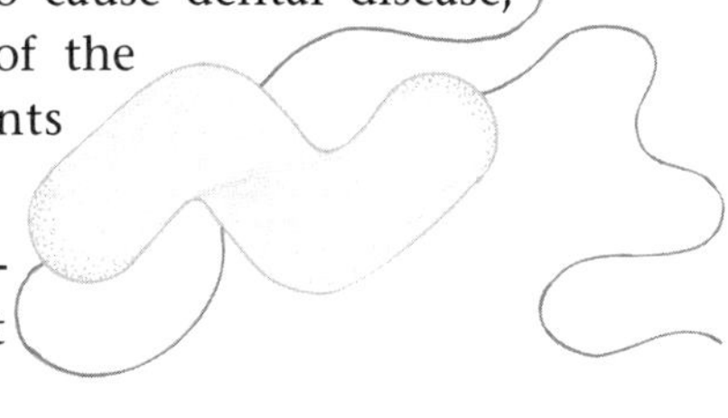

Campylobacter gastroenteritis is the most

common gastroenteritis diagnosed in the United States. It is contracted from contaminated food, milk, and water. It is seldom transmitted person-to-person, and does not multiply in food as do many other bacterial pathogens.

Helicobacter pylori

H. pylori is a Gram-negative rod that is very similar morphologically to *Campylobacter*. It colonizes the stomachs of many people at early ages. By the age of 60, nearly 50 percent are infected. This organism was only recently discovered, and the extent of its pathogenicity is still not known. Possible modes of transmission are oral–oral, fecal–oral, and contact with contaminated environmental sources, such as water.

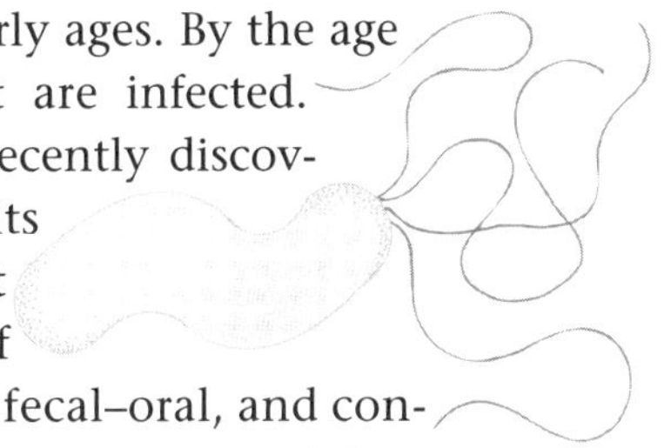

H. pylori may cause peptic ulcers and stomach cancer in people colonized by it. Discovery of its role in these illnesses has greatly changed medical attitudes, much to the benefit of patients. Stomach ulcers, once blamed on nervousness and stress, are now often cured with antibiotic treatment aimed at *H. pylori*.

This bacteria is unusual in its ability to tolerate the stomach's strongly acidic environment. It converts urea into ammonia, which helps make a less acidic niche in which it can survive.

Yersinia entercolitica

Y. entercolitica is a Gram-negative rod. This member of the family Enterobacteriaceae is a true pathogen that causes intense gastroenteritis. Several common domestic animals host *Y. entercolitica*. Yersiniosis is contracted by eating incompletely cooked pork and consuming dairy products. This pathogen may also be contracted by ingesting contaminated drinking water.

Legionella pneumophila

L. pneumophila is a thin, Gram-negative rod that stains very faintly with the Gram stain. It is widely distributed in the environment, especially in warm water. Rivers, lakes, and soil all harbor *L. pneumophila,* and its proclivity for warm water makes it common in air-conditioner, cooling tower, and humidifier water. Whirlpools and medical equipment may also contain this persistent organism. *L. pneumophila* is not transmitted by person-to-person contact. It reaches its host's lungs via aerosols, often those used in cooling equipment, humidification units, and so forth.

L. pneumophila causes the respiratory disease legionellosis, sometimes called Legionnaires' disease. While attending an American Legion convention at a Philadelphia hotel in 1996, several people fell victim to this severe illness and hospital laboratories could not identify the cause. Despite the administration of antibiotics typically used to treat pneumonia, 34 of the 221 afflicted people died.

Months later, researchers at the Centers for Disease Control finally managed to isolate a bacteria later named *Legionella pneumophila*. Further study found that the organism is most susceptible to erythromycin, an antibiotic that is not routinely used for pneumonia treatment. This information now allows for more effective treatment of patients with legionellosis, provided that proper diagnosis identifies the disease. Laboratories now can culture and grow *L. pneumophila* and perform other tests to help identify the disease.

Pseudomonas aeruginosa

P. aeruginosa is a small rod that stains Gram-negative. It is distinctly different physiologically from the Gram-negative rods of the family Enterobacteriaceae. This opportunistic pathogen is ubiquitous in nature. It can withstand and grow at a variety of temperatures, and it is resistant to several antibiotics. In particular, *P. aeruginosa* thrives in an oxygen-rich environment.

Healthy people can come in contact with *P. aeruginosa* and even ingest it in small amounts without experiencing any ill effects. Like other opportunistic pathogens, it requires compromised host defenses, such as might arise from an eye injury, skin abrasion, or burn, or prolonged contacts, such as water lodged in the host's ear canal. Invasion is also aided by foreign objects present in the body, such as indwelling urinary catheters or tracheal tubes. Young patients with cystic fibrosis (a genetic lung disorder) are especially susceptible to *P. aeruginosa* and related bacteria. These children are unable to clear the bacteria from their impaired lungs, requiring frequent respiratory therapy.

Water treatment facilities normally are not concerned with *P. aeruginosa*, but the bacteria is a priority for whirlpool operators. The warm, churning, oxygenated water allows rapid chlorine dissipation and encourages *P. aeruginosa* growth. If patrons receive abrasions from a whirlpool surface, the bacteria may invade the skin, causing severe lesions. *P. aeruginosa* is resistant to many antibiotics, requiring clinical treatment with powerful and sometimes toxic antibiotics.

Clostridium perfringens

C. perfringens is an anaerobic, Gram-positive rod and is a normal inhabitant of the human gastrointestinal tract.

C. perfringens rods are capable of encasing their chromosomes and essential organelles into tiny packages called endospores (sometimes referred to as *spores*).

These endospores are very resistant to the effects of heat, desiccation, chemicals, and other environmental factors.

Multistage water treatment generally removes *C. perfringens,* and the organism poses no threat if consumed in low numbers. This species may be a very aggressive, opportunistic pathogen in wounds, though. It produces several tissue-necrotizing toxins, which contribute to its infectious effects. Gas gangrene is the hallmark disease of this bacteria.

This bacteria also produces two food-poisoning toxins. With a doubling rate of 6 minutes, this species can rapidly grow to significant levels in foods held at room temperature. This possibly is especially significant, because its heat-resistant endospores can survive most cooking temperatures.

Because *C. perfringens* endospores have approximately the same resistance to common water treatment processes as oocysts of *Cryptosporidium parvum,* the water industry can use the endospores as an indicator of oocyst removal capabilities. These organisms may someday also become reliable indicators of waste treatment effectiveness in removing fecal contaminants because of their common presence in human waste.

C. perfringens' close relative *Clostridium botulinum* is the infectious agent of botulism. Similar in its temperature resistance, its toxin is extremely powerful and deadly to ingest.

Mycobacterium avium-intracellulare

The species *M. avium-intracellulare* is a member of the same genus as the bacterium that causes tuberculosis. Unlike *Mycobacterium tuberculosis, M. avium-intracellulare* is not contagious and has low pathogenicity for the average healthy person. Microscopic observation requires staining with a special acid-fast stain.

M. avium-intracellulare has been isolated from finished water in distribution mains. It is ubiquitous in the natural environment and is carried by numerous domestic animals.

Immunocompromised people suffer most from infection by this opportunistic organism. In this group, it becomes a respiratory and systemic pathogen. Up to 50 percent of AIDS patients have disseminated *M. avium-intracellulare* disease upon their deaths.

Aggressive water treatment for this bacteria is not presently encouraged. Lakes and rivers seem to be its primary reservoirs.

Staphylococcus Species

Staphylococci are Gram-positive cocci. They are arranged in clusters when viewed microscopically.

S. epidermidis heavily colonizes the skin of most people. It performs as an opportunistic pathogen in conjunction with surgical stitches, skin grafts, prosthetic heart valves, and so forth.

Its only significance in the water industry comes from its common presence as a sampling faucet contaminant or as a biofilm in piping. Cultures of distribution water samples, collected to monitor water system quality, may yield large counts of *S. epidermidis*. This result likely comes from people touching the sampling faucet. Biofilms spoil sampling lines and deplete residual chlorine levels.

S. aureus occasionally colonizes human skin. It commonly infects wounds and sometimes becomes a very invasive pathogen. Clinical treatment is performed with antibiotics, and *S. aureus* is becoming resistant to many of them.

S. aureus may appear in cultures of water samples as does *S. epidermidis*. Otherwise, this species is seldom significant for the water industry. However, concern is growing over *S. aureus* and other skin-infecting organisms as new attitudes toward the safety of water require more than potability.

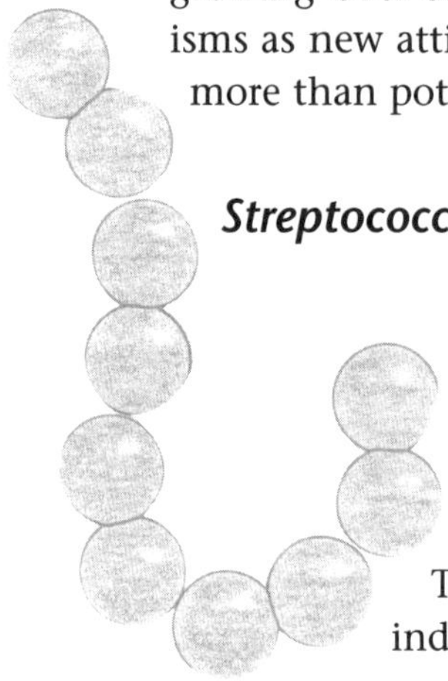

Streptococcus Species

Streptococci are Gram-positive cocci that form chains. A variety of *Streptococcus* species colonize the skin, mouths, and intestines of humans, and some produce significant infections. They cause no trouble for the water industry, but may be detected in sampling.

Enterococcus faecalis

E. faecalis is a Gram-positive cocci that arranges in pairs and short chains. This species was once considered part of the *Streptococcus* genus. *E. faecalis* is very common in human and animal intestines. It has been used as an indicator of fecal contamination of source water, although the coliform group now has priority in this role.

Bacillus Species

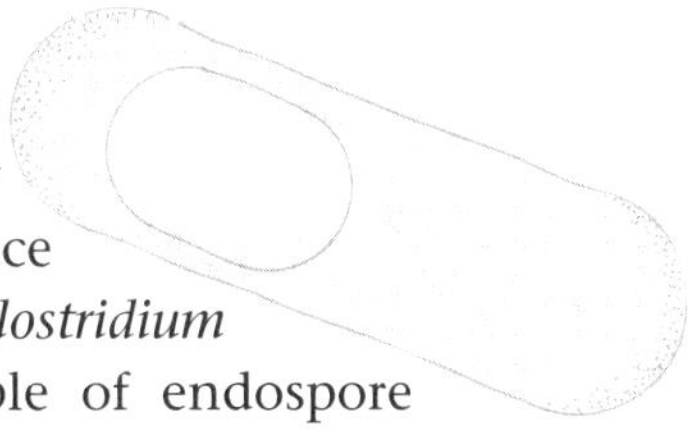

A variety of *Bacillus* species are commonly found in natural surface water and in soil. Like *Clostridium* species, they are capable of endospore production.

Most water treatment processes effectively remove or inactivate *Bacillus* species, however some cells may survive and show up in samples. The colonies are often large and sometimes spread across the surface of culture plates, obscuring the growth of other bacterial species. *Bacillus* species are harmless if consumed in low quantities, so they are not problems for the water industry.

B. anthracis causes the disease anthrax, which people may contract by handling animal hides or by inhaling or ingesting the endospores. Anthrax may become a severe, even fatal infection. This species is rare in the United States, however.

Highly bred *B. anthracis* strains are sometimes propagated by military organizations as agents of biological warfare. Its potential as a terrorist weapon is worrisome, but spreading it through drinking water systems would be a difficult task.

B. cereus and Other *Bacillus* Species

Food poisoning by *B. cereus* and a few other *Bacillus* species is fairly common, especially involving sandwich meats. The bacteria grow on the food and eventually produce enough toxins to cause illness. This group may seriously infect traumatic eye injuries. Clinical treatment includes antibiotics.

Pipe-Corrosion Bacteria

Actinomycetes

The *Actinomycetes* group of bacteria form thin, filamentous, Gram-positive rods. They are very common inhabitants of the soil, often accounting for "earthy" or musty odors through production of geosmin and other compounds. These odoriferous end products may affect the aesthetic quality of drinking water, requiring treatment with powdered activated carbon or other methods. (Geosmin is also produced by some algal species.)

Actinomycetes also colonize sand filters and sludge beds. Filter flushing and accelerated sludge bed removal helps to limit their presence and odor production.

Streptomyces is a common genus of the *Actinomycete* group.

Iron Bacteria

The iron bacteria group includes several genera: *Leptothrix, Thiobacillus, Clonothrix, Sphaerotilus, Hyphomicrobium, Caulobacter,* and *Gallionella*. These genera are widely distributed in water and soil.

They are very different from infectious bacteria such as coliforms. Iron bacteria derive energy by converting soluble (dissolved) iron into insoluble iron compounds. Some genera also precipitate manganese. When these bacteria encounter sufficient supply of iron in well or utility water, they can produce large amounts of insoluble end products. The water becomes fouled, producing taste-and-odor problems. Human disease is not a direct consequence.

Water mains may accumulate iron bacteria and related end products. This biofilm decreases residual chlorine levels and reduces water flow. It also allows other bacteria to grow, producing pipe corrosion. Some wells may become so overwhelmed with the growth of iron bacteria that the pumps and well shafts become plugged with copious amounts of thick slimy growth.

Water with small amounts of dissolved iron or manganese allows limited iron bacteria growth. Properly designed wells and water systems also help to discourage the growth of the bacteria. When iron bacteria are present in water rich in iron and manganese, continuous chlorination helps restrict their growth.

Sulfur Bacteria

The sulfur bacteria group includes several genera: *Desulfovibrio, Deulfotomaculum, Desulfomonas, Thiobacillus, Beggiatoa, Thiothrix, Chlorobium,* and *Chromatium.* These anaerobic bacteria reduce some soluble sulfur compounds to hydrogen sulfide (H_2S) gas, giving water a rotten-egg smell.

Sulfur bacteria are often found in tubercles along with other bacteria. Tubercles are swollen, corroded spots on the interior of pipes. They may enlarge and corrode the metal until cavitation and perforation occur.

Sulphur bacteria may occur in combination with other genera of bacteria that produce slime, which helps protect them both from residual chlorine. Slime also helps to retain metabolic end products produced by assimilation of organic material. These end products complement the corrosion process.

Correction of damage to water systems by sulfur bacteria is a difficult problem. Only intense chlorination can overcome the slime and tubercle barriers. High dissolved oxygen levels help to prevent growth of these anaerobes.

Nitrifying Bacteria

The group of nitrifying bacteria includes *Nitrosomonas, Nitrobacter, Nitrosovibrio, Nitrosococcus,* and *Nitrospira*. These genera are common soil bacteria that derive their energy from converting ammonia and other nitrogen compounds to compounds such as nitrite and nitrate, an essential step in nature's nitrogen fixation cycle. Plants then assimilate the nitrate compounds.

Ammonia applied as agricultural fertilizer is oxidized into nitrate compounds by nitrifying bacteria. If these nitrate compounds are washed or drained into waterways before being assimilated by crops, they become chemical (specifically, nutrient) contaminants in source waters.

Nitrate compounds may be produced from otherwise organically rich soils, also. Erosion of these soils adds nitrates to waterways, but good land-use practices can prevent the problems.

Nitrifying bacteria can become a nuisance during water treatment. Water systems that use chloramines as disinfectants may experience an increase in the breakdown of chloramines, because of the nitrite generated by some genera of these organisms. Nitrite also raises the chlorine demand in a system that relies on chlorine for disinfection.

Cyanobacteria (Blue-Green Algae)

Cyanobacteria is a group of bacteria with many naturally ubiquitous genera that resemble algae. A few examples are *Oscillatoria, Anabaena, Gleotrichia,* and *Microcystis*. Cyanobacteria perform photosynthesis as do algae; however, their cellular and physiological characteristics more closely resemble bacteria.

Cyanobacteria are most notable in water treatment for the taste-and-odor problems they create. Source water may be treated with powdered activated carbon (PAC) to remove the odorous compounds, but cyanobacteria may colonize or proliferate in a utility's sand filters, where PAC treatment is not a feasible option. Chlorination of the filters may be required to control the problems.

If stagnant ponds grow cyanobacteria in large quantities, enough toxins may accumulate from their metabolism to poison livestock that drink from the ponds. Evidence suggests that humans would be sensitive to these toxins also, but people seldom drink from ponds that are obviously contaminated.

3 Viruses, Protozoa, and Other Organisms

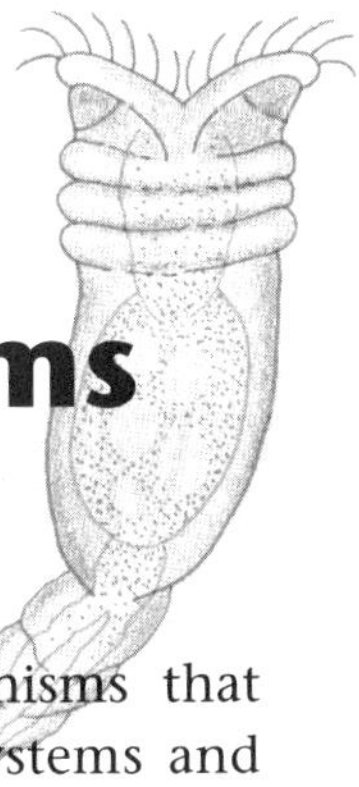

Bacteria are not the only organisms that demand attention from water treatment systems and personnel. Even smaller and less complex viruses may cause outbreaks of waterborne disease. Various species of protozoan parasites also cause illness. Algae cause taste-and-odor problems, and multicellular organisms like nematodes and rotifers pose additional challenges.

Viruses

Average size = 1/10,000 millimeter in diameter

Viruses are very simple, tiny life forms that do not multiply outside of living host cells. They can survive on their own, however, for a few minutes to several hours. They may spread by hand-to-mouth transfer, by aerosol inhalation, or by ingestion.

Virus organisms are composed of protein packages of varying designs, all with the primary function of carrying the virus chromosome from one host cell to another. Upon attaching to the host cell, a virus injects its chromosome, which then incorporates into the host chromosome, reprogramming the cell to make viruses. Once many copies of the virus are made, the cell ruptures, releasing the new viruses to scatter and infect new host cells.

These new viruses usually reinfect host cells of the same organism, but those infecting the respiratory tract may be blown into the air with a sneeze or cough, or simply through speech. Hand-to-hand and fecal–oral transmission are also possible, especially with intestinal viruses.

Viral infections are not treatable with antibiotics. They may be prevented by immunizations, careful hygiene, and effective water and sewage treatment. These methods address infection indirectly, but they have been highly successful in reducing disease, especially large-scale epidemics.

Disinfection with chlorine or ozone allows drinking water treatment facilities to easily remove most viruses from water. However, some indications suggest that viruses may reactivate (become viable again) in the distribution system, despite apparently effective treatment. Utilities that practice lime softening reduce the likelihood of reactivation because the high pH enhances the killing effect of regular treatment. Water treatment facilities that are unsure of their performance should check for viruses in water from mains distant from the point of disinfection to check for reactivation.

Hepatitis A Virus

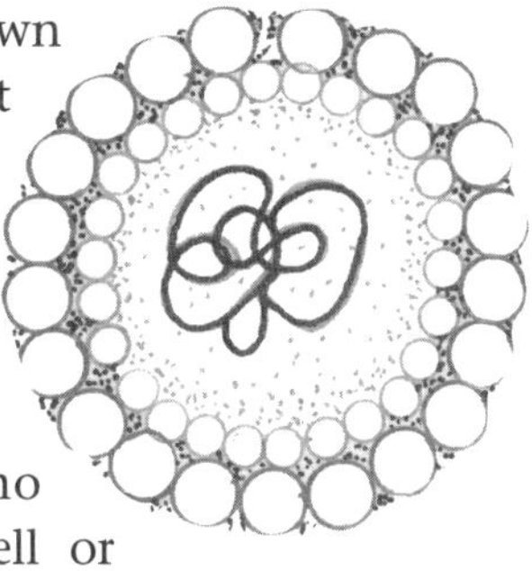

This hardy virus is also known as infectious hepatitis and is most often transmitted by fecal–oral contact. Hepatitis A may be contracted from drinking water contaminated by sewage, or by eating food that has been handled by infected people who haven't washed their hands well or taken other precautions like wearing clean food-handling gloves. Ingestion of contaminated shellfish is another common route of hepatitis A contraction. Day-care centers may be sources of outbreaks because children often exhibit poor hygiene practices.

The disease may range from mild to severe, depending on the health and age of the host. Young patients often have mild cases of the disease, while those in the middle-aged to elderly range suffer more severe symptoms.

Hepatitis viruses attack the host's liver. Symptoms of this disease include fever, nausea, muscle aches, and vomiting; the liver may also swell and cause pain.

No effective cure for hepatitis is known, and the disease resolves itself in 4 to 8 weeks. Of the few who die from the disease, 70 percent are over 49 years old.

Enteroviruses (Poliovirus, Coxsackievirus, Echovirus, and Others)

This large group of viruses occur commonly. Most of the diseases are characterized by generalized symptoms that affect many organs, including the brain. These organisms may express their pathogenicity in other forms as well, such as summer colds or stomach or intestinal flus.

A notable disease in this category is poliomyelitis, caused by the *Poliovirus*, which is now rare in the United States. An intensive nationwide immunization program has minimized the threat of transmission.

Waterborne transmission of the enteroviruses is possible, but person-to-person transmission is more common. However, as with other waterborne diseases, ineffective water treatment allows a potential for large-scale epidemics.

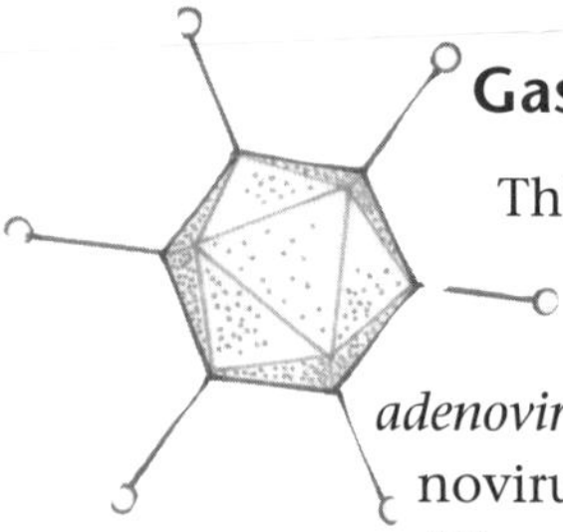

Gastroenteritis Viruses

This large group of viruses cause a variety of stomach and intestinal flu symptoms. It includes the *Norwalk virus, rotavirus,* and *adenovirus,* along with many other genera. The adenoviruses are notable because they may cause either respiratory disease or gastrointestinal disease. They are very contagious, and waterborne and person-to-person transmissions occur.

Protozoa

Average size = 1/100 millimeter in diameter

Most protozoa are larger than bacteria. They are single-celled organisms, as are viruses and bacteria, yet they possess more complex physiologies and life cycles. In particular, a protozoan cell incorporates a nucleus, which contains its chromosomal DNA.

A variety of forms of protozoa are known. In fact, they are developed to the point that generalizations about their shape and nature are difficult to make.

Paramecium

Paramecium species are present throughout nature, but they have no importance for the water industry. They are, however, classic examples of protozoa that commonly live in natural waters.

Many high school students encounter paramecia while microscopically examining pond water. The pear-shaped cells of this genus are covered with tiny hairs called *cilia* that beat rhythmically to propel the organism through the water environment. Protozoa with cilia are called *ciliates.*

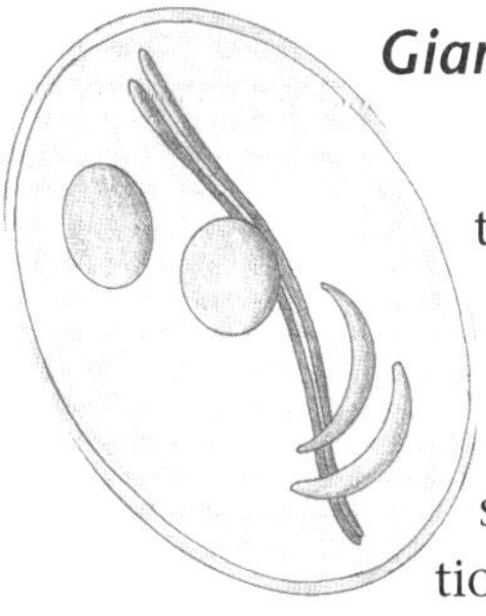

Giardia lamblia

G. lamblia has profound importance for the water industry. It is a common intestinal parasite of humans and other mammals. It exists in two stages: the active trophozoite stage and an inactive cyst stage during which it is resistant to conditions in its environment.

Beavers are notorious for passing *Giardia*, which they harbor in their intestines, via their feces into apparently clean mountain streams. People lured by visions of nature's beauty and purity may drink from the cold mountain streams and find themselves suffering from giardiasis (camper's diarrhea). This disease creates severe diarrhea, cramping, and fatigue. It is treatable with antibiotics when correctly diagnosed; however, untreated giardiasis may continue for weeks before self-resolving.

Giardia cysts are also transmitted from person-to-person via the fecal–oral route, an especially common problem in day-care centers where young children may not practice good hygiene.

The inactive cysts are resistant to chlorine, so they may survive treatment if they pass through a facility's filtration system and disinfectant contact time is not sufficient to overcome resistance.

Ingested *Giardia* cysts mature into trophozoites, which then multiply. Millions of organisms attach to the wall of the small intestine and cover so much of it that they interfere with nutrient absorption. Diarrhea accompanies the limited nutrient absorption, creating great fatigue and weight loss if the disease is prolonged.

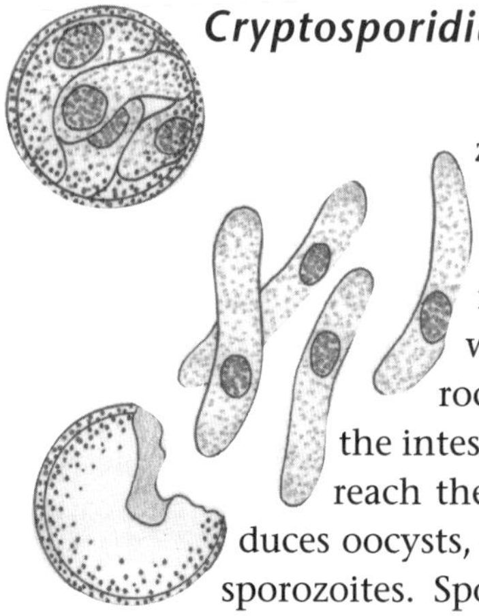

Cryptosporidium parvum

C. parvum is a species of protozoa quite different from free-living organisms such as paramecia. It is active only inside a mammalian host's small intestine where it develops into a small, rod-shaped *sporozoite* and invades the intestinal lining. It develops further to reach the reproductive stage when it produces oocysts, each containing up to four more sporozoites. Sporozoites with thin oocyst walls are immediately released to reinfect the same host's intestinal lining. Other sporozoites with thick oocyst walls are defecated into the environment or wastewater system.

The disease of cryptosporidiosis involves severe diarrhea accompanied by intestinal cramps and fatigue. The infection may last from a few days to a few months, depending on the strength of the host's immune system. Serological studies have found antibodies to *Cryptosporidium* in many people, indicating past infections that had not been attributed to cryptosporidiosis. This phenomenon likely accounts for the relatively mild cases that occur during epidemics.

People with compromised immune systems, such as leukemia patients, transplant recipients, infants, the elderly, and individuals with AIDS, are endangered most by the disease. This group may add up to one-fourth of the population. No effective antimicrobial treatment is presently available for treatment of cryptosporidiosis. Person-to-person transmission is possible when proper hand washing is ignored. Also, ingestion of oocysts from lakes, streams, swimming pools, or poorly treated drinking water may precipitate cryptosporidiosis. Many mammals, including cattle (especially calves), hogs, and humans, are susceptible to cryptosporidiosis, so they can also transmit the disease.

The largest waterborne epidemic of the disease occurred in Milwaukee, Wis., in 1993. Investigators believe that huge numbers of *Cryptosporidium* oocysts flowed into Lake Michigan in runoff from cattle stockyards or other flows of sanitary sewers following heavy rains. The plume of oocyst-laden water was apparently drawn into the city's drinking water facility, which uses lake water as its souce.

High numbers of *Cryptosporidium* oocysts can escape coagulation with conventional treatment chemicals, allowing them to pass through granular filtration systems. The oocysts are also very resistant to the killing effects of chlorine. In Milwaukee, the infective oocysts reached the distribution system and were ingested by most of the utility's 600,000 customers. An estimated 400,000 people became ill with gastroenteritis, likely caused by *C. parvum.* Up to 100 deaths occurred during the outbreak involving primarily leukemic children and other individuals with weakened immune systems.

Cyclospora cayetanensis

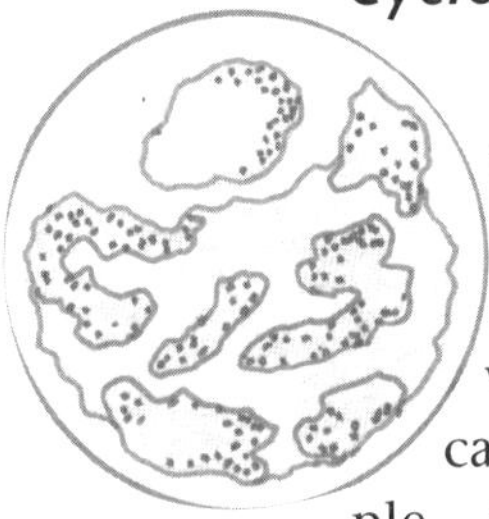

C. cayetanensis is a protozoan species that is similar to *Cryptosporidium* in its structure and in the disease that it causes. It has been implicated in a few waterborne disease outbreaks. It also has caused food-borne epidemics when people ingested contaminated raspberries imported to the United States from Central America. The common antibiotic Trimethoprim/Sulfamethoxazole is effective for treatment.

Microsporidia Group

This group of protozoa—mainly the genera *Encephalitozoon, Nosema, Vittaforma, Pleistophora, Enterocytozoon,* and *Microsporidium*—infects both animals and humans. These parasites invade the intestinal lining where they proliferate into high numbers, causing the host to experience intestinal illness.

The microsporidia produce thick-walled spores that are resistant to environmental stresses. This resistance allows them to pass from infected individuals to new, distant hosts, via water or soil. Host-to-host transmission is also likely.

The significance of microsporidia for the water industry is still not fully defined. The prevalence, treatment resistance, and pathogenicity of these genera are presently under study.

Detection of the tiny spores (2–3 µm in diameter) is relatively difficult. Microsporidiosis is a disease with no present antibiotic treatment.

Amoeba

Water treatment personnel must deal with many amoebae, especially *Acanthomoeba, Entamoeba,* and *Naegleria*. The flexible cell walls of these organisms allow them to "ooze" in one direction or another by sending out projections called *pseudopodia*. The rest of the cell's cytoplasm follows into the pseudopodia until the entire cell has been transported. Amoebae make cysts that are resistant to environmental stresses such as drying, which helps the species survive and reach new niches.

Entamoeba histolytica is an intestinal pathogen. This invasive amoeba is contracted by person-to-person contact or by ingestion of water contaminated with feces or foods irrigated with contaminated water. When treated wastewater is used to irrigate vegetable crops, such as lettuce, cysts may be trapped in the plants' crevasses. If the vegetables are not washed well, the cysts may be ingested and continue their life cycle.

E. histolytica invades and multiplies in the host's intestinal tissue. If the disease is protracted, the amoebae may travel to the liver or brain causing development of amoebic abscesses. Antimicrobial treatment becomes difficult once the infection has reached this stage.

Acanthomoeba, Naegleria fowleri, and *Balamuthia* are free-living amoebae that inhabit soil, ponds, and rivers. They do not require an animal or human host to survive. *Acanthomoeba* and *Balamuthia* are rare yet particularly notorious pathogens. Each can cause severe encephalitis, a brain infection that can progress rapidly, usually with fatal results. Swimmers in ponds and lakes harboring these organisms may contract them through the nasal passages. The amoebae progress to the nasal olfactory lobes and eventually to the brain.

More commonly, *Acanthomoeba* causes infections of the ears, maxillary sinuses, and lungs, and especially keratitis of the eyes. Keratitis occurs when *Acanthomoeba* invades the cornea, often after contact lenses are exposed to contaminated water and then worn for an extended time. Extended exposure gives the amoeba an opportunity to create a corneal ulcer. People that prepare their own saline washes for contact lenses and those who swim while wearing their lenses have the greatest risk of infection. Corneal ulcers are painful and must be aggressively treated to prevent thorough destruction of the eye.

The amoeba *N. fowleri* is a free-living soil and water organism. It can invade the human nasal chamber and eventually the brain, causing severe meningoencephalitis.

Algae

Average size = 1/100 millimeter in diameter

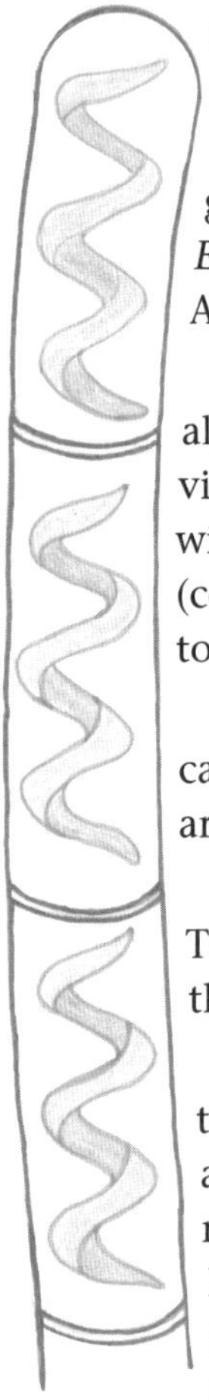

The surface waters of the world are filled with algae. Numerous species are included in several groups of green, yellow-green and golden-brown, Euglenoid, and Cryptomonad algae.

Numerous excellent color drawings of algal groups can be found in *Standard Methods for the Examination of Water and Wastewater* (APHA, AWWA, WEF, 1998).

Apart from their photosynthetic process, many algae are very much like protozoa. Some exist individually (as unicellular organisms), perhaps even with flagella for motility. Others grow in groups (colonial organisms), and yet others exhibit filamentous or tubular structures.

Algae that remain freely suspended in water are called *planktonic algae*. Those that attach to surfaces are called *sessile algae*.

Diatoms are algae with cell walls of hard silica. They may be loosely viewed as "glass seashells" of the microscopic world.

Dinoflagellates may threaten fish, or cause "red-tide" conditions that affect fishing and recreational activities. One such organism, *Pfiesteria*, produces neurotoxins that cause lesions and death in fish. Humans may also develop skin lesions, respiratory problems, and central nervous system problems if they come into significant contact with the toxins.

Dinoflagellates are common, normally harmless inhabitants of most rivers, lakes, and oceans. However, their role may change when excessive nutrients such as nitrates, phosphates, and sewage enter the water. The change in environmental conditions apparently allows dramatic proliferation of *Pfiesteria,* creating conditions for disease in fish and the humans that contact the water. The Menhaden fish is the primary host of *Pfiesteria.* This fish is harvested for its oil, which is then used in cattle feed and in cosmetics.

Presently, *Pfiesteria* is no threat to drinking water facilities. Nevertheless, potential for emergence as a powerful source of disease demonstrates how environmental changes can change an organism's role in nature, in surprising and worrisome ways.

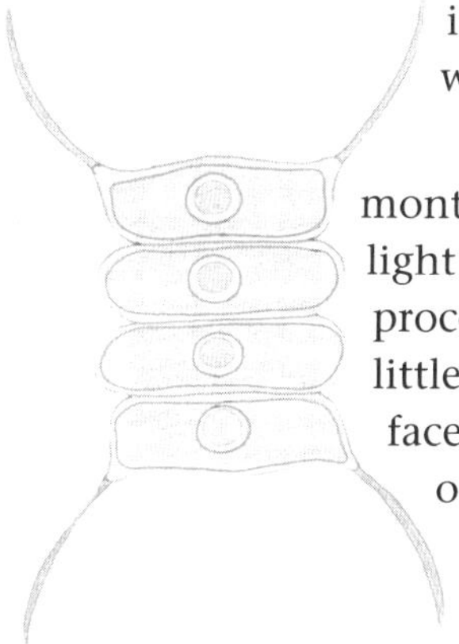

Algae flourish during warm, sunny months when nutrients are high and ample light is available to fuel photosynthetic processes. Finished water regulations make little mention of algae, but operators at surface-water facilities usually are well aware of the potential impact of these organisms on plant operations and finished water quality. Algal blooms in source waters may create taste-and-odor problems, filters may clog requiring significantly more frequent backwashing, and finished water color problems may arise from chlorophyll that is dispersed in the water from ruptured algae cells.

Furthermore, strain on initial steps in multistage systems may allow increases in organic compounds derived from algae, even at the clearwell. Consequently, chlorine demand may rise.

Algae (other than blue-green algae) play a very beneficial ecological role. They help oxygenate water and remove numerous chemical contaminants such as phosphorous, ammonium, and nitrate compounds. They are an essential part of complex food chains, giving them a vital role in maintaining the overall health of a lake or river.

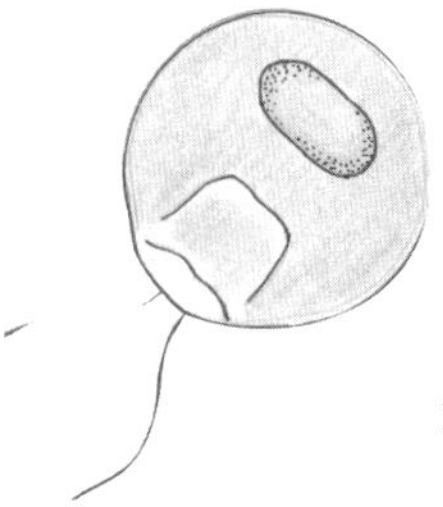

Algal blooms in a reservoir can be controlled by dispersing copper sulfate in the water. When properly administered, this practice is effective, but planning must account for the toxic effect of copper sulfate on snails and other water animals and plants that are important in maintaining a healthy reservoir. The chemical should be used at careful dosages.

Multicellular Organisms

Water treatment facilities must also contend with small multicellular organisms. These organisms are not pathogenic for humans, but they may interfere with treatment processes.

Nematodes are microscopic round worms that are common soil and water organisms. They may find ideal habitats in the sludge blanket of a treatment basin, or the mixed media matrix of a rapid filter. Their numbers may rise to the point that they cause filter clogging. Because of their impressive resistance to chlorination, they may also reach finished water. Customers with acute vision may spot the worms in their drinking glasses, compromising water aesthetics. If nematodes from source water survive treatment and pass into the finished water, they may carry along viable pathogenic bacteria that they have ingested.

Rotifers are microscopic creatures that flourish in water high in nutrients. They are distinguished by circles of cilia around their mouths. The movement of the cilia resembling revolving wheels directs food particles into their mouths. The water industry seldom has a problem with rotifers. Particle analyses on high-nutrient or otherwise polluted water may find a large number of rotifers, however.

Small crustaceans and *insect larvae* may survive treatment to reach finished water at some facilities. This possibility is especially likely when water is unfiltered or when mains and filters become colonized with the organisms.

These tiny organisms do not cause disease; however, their presence in drinking water is clearly undesirable. They also can disrupt filter operation, which may raise finished water turbidity.

4 Laboratory Methods for Isolation and Detection of Pathogens

To fulfill their critical role in safeguarding public health, personnel at water treatment facilities carefully monitor water quality by conducting a number of tests designed to isolate and detect pathogens. The federal Safe Drinking Water Act defines acceptable practices for sampling and analysis of drinking water. The law also grants regulatory primacy to state agencies, so their regulations take precedence, provided the states impose requirements equivalent to or more stringent than those of the SDWA. Primacy allows states to address specific needs that do not appear in other locations while ensuring effective protection for public health.

As a result, this handbook cannot possibly summarize all the requirements for collecting and analyzing water samples in a particular location. Instead, it gives general information that may be supplemented by specific guidance from local regulators.

Also, detailed procedures for conducting specific tests are beyond the scope of this book. More detailed information is available in *Standard Methods for the Examination of Water and Wastewater* (APHA, AWWA, WEF, 1998). Useful summaries for some procedures are available in AWWA M12—*Simplified Procedures for Water Examination* (AWWA, 1997). *An Operator's Guide to*

Bacteriological Testing (Lisle, 1993) provides additional explanation and background information in an accessible question-and-answer format.

Sampling Methods

Proper sample collection and handling is the most important consideration, and often the least discussed, in any monitoring program. The goal of any sampling procedure is to gather representative samples that accurately reflect conditions, especially the presence of pathogens, in the water being tested. Test results are only as reliable as the techniques for collecting samples.

A public water utility must gather samples according to a written plan, subject to review and revision, that specifies sampling sites, frequency, and other details. The number of samples to be taken each month is specified based on the size of the community served by the utility. If a routine sample tests positive for certain pathogens, regulations may require collection of additional samples for followup testing.

Samples are collected in sterile bottles or plastic bags and immediately dosed with sodium thiosulfate ($Na_2S_2O_3{\bullet}5H_2O$). Sample containers must be kept closed until the moment of sample collection. Sampling should be avoided on windy and rainy days, if possible, because these conditions increase the probability of contamination. The container's lid should not be placed on any surface or held face up, which could also increase the risk of contamination. The sample should be refrigerated or iced and sent as rapidly as possible to a certified laboratory for analysis. Additional guidelines may apply, depending on conditions such as water temperature and the source of the sample—a laboratory sampling tap, source location (lake, river, or impoundment).

Methods of Testing for Bacterial Pathogens

Federal and state agencies require testing of drinking water samples for total coliform bacteria. Although these organisms generally are not harmful themselves, as discussed in chapter 2, the presence of coliforms is often associated with contamination by more actively pathogenic organisms. Therefore, they are accepted as an index of the microbiological safety of drinking water.

For this reason, the US Environmental Protection Agency (USEPA) has promulgated the Total Coliform Rule, which requires public notification and corrective action if 5 percent or more of a water utility's samples test positive for total coliforms. Different thresholds for numbers of positive samples apply for small drinking water systems. In addition, any positive result in a total coliform test requires further sampling and retesting. If the new samples also test positive, additional testing for fecal coliforms and *Escherichia coli* is required. If these contaminants are present, they pose a serious health risk requiring a rapid response, including notification of regulators and the public.

Conventional Testing Methods

Some methods for conducting bacteriological tests involve quantitative counts to determine the most probable number (MPN) of organisms in a sample. Other methods determine only the presence or absence of organisms without attempting to quantify them. All of these methods involve combining water from the sample with liquid or solid nutrient media. The media encourage growth of coliforms while suppressing growth of other bacteria.

A solid medium, or *agar*, forms a gel in the testing plate. It may include a variety of ingredients, depending on the group of bacteria that the microbiologist wants to grow. Bacteria grow in colonies on the agar surface. Colonies are separate, countable groups of bacterial cells that have theoretically developed from one cell growing at a logarithmic pace. Colonies may be picked off with a wire needle or loop and transferred to a special biochemical test media, where they react to chemicals during incubation. Patterns of their reaction are noted, and a database is used to identify them.

A liquid medium is called a *broth*. Broths of varying composition are used to grow specific types of bacteria. If organisms are present, they multiply during incubation and diffuse throughout the medium, producing a cloudy or turbid appearance. Samples are then examined under normal light and ultraviolet (UV) light. Most media are designed to inhibit the growth of unwanted species of bacteria while nourishing the growth of the species that is being tested. In addition, the chemical composition of the medium causes colonies of selected species to appear in a particular color. These media are called *selective* or *differential* media. Additional biochemical tests may be performed to more accurately identify the colonies of bacteria.

One test of this type is the *ONPG-MUG test* (named for the chemicals in the medium, which include ortho-nitrophenyl-β-D-galactopyranoside, or ONPG, and 4-methylumbelliferyl-β-D-glucuronide, or MUG). This test is sometimes called the *MMO-MUG test,* in which *MMO* stands for *minimal media ONPG.*

Colilert™—A Presence–Absence or MPN Test

Colilert™ is a proprietary system for the ONPG-MUG test approved by USEPA for analysis of total coliform and *E. coli* in drinking water. The media, which includes lactose, is added to a water sample. Coliform bacteria metabolize lactose, releasing compounds that react with a chemical in the medium to produce a yellow color after 24 hours of incubation. Therefore, such a color change indicates the presence of any of four primary coliform genera. The test can specifically detect *E. coli,* as well, because that species generates a compound that produces a bright blue fluorescence in the medium under UV light.

The Colilert™ test can also generate quantitative estimates of bacteria count. In this variation, the water–medium mixture is sealed in a card-like tray with several wells or in multiple tubes. The number of wells or tubes that show positive presence of coliforms is compared to an appropriate MPN index table to evaluate the extent of contamination in the sample. (See Table 9222:II in *Standard Methods for the Examination of Water and Wastewater.*)

Colisure™ Media

The Colisure™ method is another proprietary system approved by USEPA for analysis of total coliform and *E. coli* in drinking water. Total coliforms and *E. coli* are assessed after the water sample is mixed with a special medium and incubated for 24 hours. The medium turns to a red or purple color when total coliforms grow. Fluorescence under UV light indicates the presence of *E. coli.*

Membrane Filtration

Membrane filtration is a testing method used by drinking water and wastewater facilities to concentrate and retrieve low numbers of bacteria from relatively large amounts of water. The test can theoretically detect a single bacterium in a 100-mL sample.

To concentrate the sample, the water is poured into a cylindrical funnel lined with a filter and attached to a vacuum manifold. The vacuum draws water through the filter, capturing any bacteria present in the sample. The filter is then removed with forceps and placed into a petri dish containing a special medium, which nourishes the growth of any organisms. The dish is incubated for 24 hours at 35°C, during which time each bacterium multiplies into a colony of billions.

Each colony theoretically represents a single organism in the original sample. However, bacteria may land on the filter in small clumps. To allow for this possibility, clumps of bacteria are counted as colony-forming units to express the amount of bacterial recovery from a sample.

Heterotrophic Plate Count

Another test method called the *heterotrophic plate count* (HPC) allows enumeration of the bacteria present in a water sample. One of a variety of nutrient agars is prepared and kept warm enough to prevent it from solidifying. Next, 1 mL of the water sample is added to an empty, sterilized petri dish, and the warm agar is poured onto it. After briefly swirling the dish to cause mixing, the agar is allowed to cool and harden. After incubation for 48 hours, the resulting colonies indicate the total count of bacteria in the sample.

Testing for Protozoa

Water utility staff must conduct more complex analysis of water samples to determine the presence of protozoa such as *Cryptosporidium parvum* and *Giardia lamblia*. Most testing of this type involves microscopic analysis to identify specific microorganisms based on their appearance and cellular structures. Cartridge filters are available that trap protozoans in samples of water passed through them. Properly used, these devices can speed up the process of analysis and detection, but a certified lab must complete the testing.

All methods of testing for protozoa require concentrations of large amounts of water into manageable volumes to confine any organisms present. Any solids in the remaining sample are further concentrated using a centrifuge. This is the most labor-intensive aspect of the testing process.

Convenient techniques for detecting protozoans are under development, and new procedures are continuously being researched. One creative method approved by USEPA is immunomagnetic separation. This procedure tags organisms with magnetite and extracts them from concentrated debris using a strong magnet. The increase in efficiency reduces the number of purification steps and thus saves time and greatly improves the recovery of organisms.

Testing for Viruses

Viruses are obligate, intracellular parasites; they can grow only inside animal or plant cells. That requirement adds difficulty for monitoring and detection.

Any testing for human viruses involves adding concentrated samples to small, flat-sided flasks prepared with thin monolayers of tissue cells. These cells are often taken from malignant growths of cancer patients and processed for commercial sale. Through microscopic analysis, specially trained personnel look for distinctive types of cellular damage unique to certain viral genera or groups. This damage is called the virus's *cytopathic effect* or CPE.

Methods of Testing for Other Organisms

Water industry personnel are seldom most concerned with fungi. These ubiquitous organisms can grow in adequately moist and warm environments with very few nutrients. Most laboratory tests for these organisms attempt to grow the fungi on agar plates and then study the resulting growths under the microscope. Because of their increasing clinical importance, molecular biology techniques are being developed to identify fungi.

Advances in Molecular Biology

The science of molecular biology has developed techniques for detecting certain species and groups of microorganisms by identifying their deoxyribonucleic acid (DNA). DNA probes detect specific chromosomal fragments known to be unique to specific organisms. Some of these methods are simple to use, while others are expensive and time consuming.

These analytical techniques hold great promise for improving detection and identification of particular pathogens. Most laboratories are waiting, however, until more streamlined and less expensive molecular biology methods become available before replacing their conventional techniques of culturing samples in nutrient media. Some companies have developed high-tech instruments; however, the instruments can be used only after laboratory processing or culturing of samples.

Light Microscopy

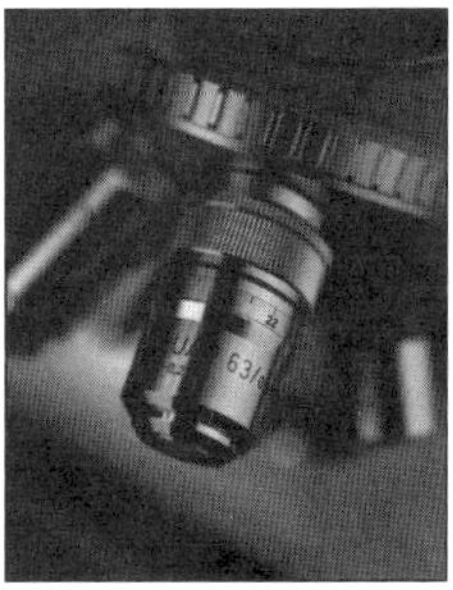

Light microscopes have long played a powerful role in studies of microbiological processes. Despite continuing innovation in analytical methods, they remain essential tools for discovery and sample analysis.

Biochemical, metabolic, and molecular biological principles can often identify characteristics of organisms far beyond what may be observed with a microscope. Yet today's sophisticated instruments retain essential advantages as well. Speed and ease of use are strengths of microscopic analysis. This familiar method remains a reliable way to confirm the presence of certain organisms, crystals, and miscellaneous organic particles.

Antonie van Leeuwenhoek used his new microscope to make the first discovery of microorganisms in 1674. Prior to that time, mites were the smallest known creatures. His discovery was the beginning of a slow emergence of the understanding that diseases could be caused by microorganisms.

Leeuwenhoek began thinking about the microscope when he noticed merchants using magnifying glasses to judge the quality of cloth before purchasing. This practice aroused his interest and inspired him to build miniature microscopes with tiny, highly refined lenses ground painstakingly by hand. Some lenses were ground with single grains of sand. Leeuwenhoek also extracted metal from ore and melted and shaped the metal himself. He produced 400 microscopes by the time he died at 91 years of age.

Using these instruments, he studied a variety of materials from human blood to elephant tusks. His discovery of bacteria and protozoa, using lenses as small as a pinhead, proved to be his greatest observations. Leeuwenhoek enthusiastically reported his findings throughout his life to the Royal Society of London, but neither he nor the others made a connection between microbes and diseases. Illnesses that impaired and killed huge numbers of people were still thought to be caused by a large assortment of evil vapors and unclean thoughts.

Finally in 1876, 200 years after Leeuwenhoek's discovery, Robert Koch and Louis Pasteur made the connection between microorganisms and disease, establishing the concept known as the *germ theory*. This leap in understanding allowed them and others to launch an attack on diseases, which has greatly improved human health and life expectancy.

The design of microscopes progressed throughout the 20th century. Monocular instruments made of brass were popular in the early 1900s, while more advanced binocular instruments were developed in the 1950s. Today, the technology of microscopy continues to advance with the addition of numerous lenses, filters, and fine adjustments. Some have evolved into complete computerized workstations.

Particle analysis of water is a great example of the value of microscopy in evaluating the varying and unpredictable characteristics of water samples. It allows definite identification of microscopic organisms and particles present in a sample. Careful counts can yield quantitative results. Few other methods can supply such rapid and practical guidance for water treatment decisions.

Alternative test methods have indirect results compared to microscopy. Most tests look for signs associated with certain pathogens and then infer the presence of the organisms. As this link becomes more indirect, the room for false-positive and false-negative results increases. Sometimes the presence of a specific pathogen may never be directly confirmed. Even more significantly, individual tests may detect only those organisms specifically targeted and may not reveal all organisms in the sample.

Measurements and Data Accuracy

All analytical methods are limited by the inherent relevance and maximum accuracy. These characteristics affect the reliability of testing from the sampling stage to the final result. Data generated from water sample studies must be expressed using these concepts.

Data relevance is an expression of how effectively an operator can use test results. Data should not be reported at accuracy levels more specific than users of the information can practically apply. For example, results from a daily test on a water component might fluctuate several whole units every hour (e.g., rising from 60 to 70 units). The report of that test need not record a result more specific than a whole number (e.g., the report might state 65 rather than 65.32). Even if a laboratory instrument accurately gives the result to the hundredths place, the reading should be rounded to the nearest whole number because the additional precision cannot improve practical control of treatment.

Data accuracy is as important as its relevance to treatment decisions. It depends on specific practices for sample collection and measurement. Testing cannot generate a final result more accurate than the least accurate measurement made during the study. When measuring a sample, the smallest pipette that can hold the entire volume should be used. A test result loses accuracy in proportion to the number of measurements taken of each sample.

When a sample is measured at 100 mL with a graduated cylinder having an accuracy of plus or minus 5 percent, the true volume lies between 95 mL and 105 mL. If the entire sample were cultured for bacteria, a microbiologist should not be overly concerned whether the colony count is, for example, 42 or 43. The limited accuracy of the initial sample measurement prevents any greater accuracy of results.

When different values are added together, the final result must be reported with the number of decimal places of the least accurate value. For example, the series 2.30 + 4.322 + 3.0 + 2.1213 equals 11.7433. The sum should be reported as 11.7.

When different values are multiplied, the result should be expressed relative to the least significant numbers in the factors. Refer to the chapter on Significant Figures in *Standard Methods for the Examination of Water and Wastewater* (APHA, AWWA, and WEF, 1998) for a detailed discussion.

Round off numbers with final digits ranging from 1 to 4 downward. Round off numbers with final digits ranging from 5 to 9 upward. For example, 3.71 through 3.74 round to 3.7 in a report listing values to one decimal place; 3.75 through 3.79 round to 3.8.

If a number is generated with a particular accuracy, any report should record it to the correct number of decimal places, even if the last digit is zero. This practice indicates to anyone reading the report that testing was performed to the indicated accuracy. For example, record 3.20 instead of 3.2 for values carried to two decimal places.

Calculators and computers often allow users to automatically round values to the nearest nonzero number. This adjustment is sometimes undesirable, and settings should be adjusted to give values at the appropriate level of accuracy.

Data entry and reporting should also avoid hanging decimal points. This condition occurs when a decimal number less than 1 is recorded without a zero preceding the decimal point. The absence of a whole number increases the likelihood of data being misread. For example, rather than recording .235, record 0.235.

5 Chemistry of Microbiology and Water Treatment

The basic laws of physics define the processes of chemistry. Similarly, chemical processes define the basic functions studied in the life sciences. For this reason, the fields of chemistry and microbiology overlap in critical ways. This relationship creates a need for water facility personnel to learn something about the basic principles of chemistry. It will help them to understand the behavior of pathogenic microbes as well as the numerous disinfection methods. Knowledge of chemistry also helps to explain how the functions of beneficial microorganisms contribute to a healthy ecosystem.

This chapter provides an introduction to the elementary concepts of inorganic and organic chemistry. It also outlines the natural process of nitrogen fixation in the environment. The last sections of the chapter discuss the chemistry of water softening and the disinfectants used by water facilities. For more detailed instruction on these topics, consult the water treatment manuals listed in the references or refer to general chemistry texts.

Inorganic Chemistry

Inorganic chemistry is based on a basic set of elements. These chemical elements are composed of atoms, which are the smallest particles that retain an element's original characteristics. Atoms can be broken down further into subatomic particles—the proton, neutron, and electron—but these building blocks of matter are stable only as parts of atoms.

Protons carry positive electrical charges. They combine with neutrons, which have no charge, to form the nucleus of an atom. Negatively charged electrons occupy the space surrounding the nucleus in a series of "shells" that surround the atom. If the atom were the size of a football stadium, the nucleus would be no larger than a small insect flying in the middle.

The defining characteristic of an atom of any element is the number of protons in the nucleus, called its *atomic number*. An atom is electrically stable when it has the same number of electrons as protons, so their opposite charges balance one another, and the atom itself has no charge. An electron may be removed from or added to the outer shell of an atom. This type of atom is called an *ion*. An ion with more protons than electrons has a positive charge and is called a *cation*. An ion with more electrons than protons has a negative charge and is called an *anion*. The number of electrons in an atom's outer shell determines how the element combines with others to form compounds.

Periodic Table of the Elements

Chemical elements have been classified in a highly organized chart called the periodic table of the elements (Figure 5-1). Elements are arranged numerically in the periodic table from atomic number 1 (hydrogen) to atomic number 103 (lawrencium). Each successive atom in this series is larger and heavier than the preceding one. Abbreviations of one or two letters represent each element's name, and other data often are included as well.

KEY

Atomic Mass (Weight) — 12.01115
Symbol — C
Atomic Number — 6

Atomic weights conform to the 1961 values of the Commission on Atomic Weights.

Period	IA	IIA	IIIB	IVB	VB	VIB	VIIB	VIII	VIII	VIII	IB	IIB	IIIA	IVA	VA	VIA	VIIA	0
1	1.00797 H 1																1.00797 H 1	4.0026 He 2
2	6.939 Li 3	9.0122 Be 4											10.811 B 5	12.01115 C 6	14.0067 N 7	15.9994 O 8	18.9984 F 9	20.183 Ne 10
3	22.9898 Na 11	24.312 Mg 12											26.9815 Al 13	28.086 Si 14	30.9738 P 15	32.064 S 16	35.453 Cl 17	39.948 Ar 18
4	39.102 K 19	40.08 Ca 20	44.956 Sc 21	47.90 Ti 22	50.942 V 23	51.996 Cr 24	54.9380 Mn 25	55.847 Fe 26	58.9332 Co 27	58.71 Ni 28	63.54 Cu 29	65.37 Zn 30	69.72 Ga 31	72.59 Ge 32	74.9216 As 33	78.96 Se 34	79.909 Br 35	83.80 Kr 36
5	85.47 Rb 37	87.62 Sr 38	88.905 Y 39	91.22 Zr 40	92.906 Nb 41	95.94 Mo 42	(99) Tc 43	101.07 Ru 44	102.905 Rh 45	106.4 Pd 46	107.870 Ag 47	112.40 Cd 48	114.82 In 49	118.69 Sn 50	121.75 Sb 51	127.60 Te 52	126.9044 I 53	131.30 Xe 54
6	132.905 Cs 55	137.34 Ba 56	138.91 La 57	178.49 Hf 72	180.948 Ta 73	183.85 W 74	186.2 Re 75	190.2 Os 76	192.2 Ir 77	195.09 Pt 78	196.967 Au 79	200.59 Hg 80	204.37 Tl 81	207.19 Pb 82	208.980 Bi 83	(210) Po 84	(210) At 85	(222) Rn 86
7	(223) Fr 87	(226) Ra 88	(227) Ac 89															

Transition Elements (groups IIIB through IIB)

Numbers in parentheses are mass numbers of most stable or most common isotope.

Lanthanide Series

140.12 Ce 58	140.907 Pr 59	144.24 Nd 60	(147) Pm 61	150.35 Sm 62	151.96 Eu 63	157.25 Gd 64	158.924 Tb 65	162.50 Dy 66	164.930 Ho 67	167.26 Er 68	168.934 Tm 69	173.04 Yb 70	174.97 Lu 71

Actinide Series

232.038 Th 90	(231) Pa 91	238.03 U 92	(237) Np 93	(242) Pu 94	(243) Am 95	(247) Cm 96	(247) Bk 97	(249) Cf 98	(254) Es 99	(253) Fm 100	(256) Md 101	(254) No 102	(257) Lr 103

Figure 5-1 Periodic Table of the Elements

In particular, each element's atomic weight is listed. This number represents the average sum of protons and neutrons in the atom's nucleus. The number of protons always remains the same for any element, but the number of neutrons may vary. Atomic weight indicates the most likely composition of the nucleus. This value is useful for comparison with other atomic weights, but it is not a measure of mass.

The elements are grouped in the periodic table relative to their general characteristics and tendencies to react with other elements. Elements in each horizontal row, called a *period*, have the same number of electron shells. Hydrogen and helium, the only members of the first period, each have only one electron shell. The elements of the second period—lithium through neon—all have two electron shells, and so on.

The vertical columns in the periodic table define the groups that share similar chemical properties. The groups are noble gases, metals, transitional metals, nonmetals, alkali metals, alkali earths, rare earths, and halogens. For example, chlorine, bromine, and iodine, all of which can be used as disinfectants in water treatment, are members of the halogen group. The elements of this group carry a negative charge and react well with positively charged alkali metal and alkali earth elements, such as sodium, potassium, magnesium, and calcium. These reactions are called *ionic bonding*, and they result in common compounds such as sodium chloride, magnesium chloride, and calcium bromide.

Chemical Compounds

A group of chemically bonded atoms forms a particle called a *molecule*. Atoms of two or more different elements can combine in definite proportions to form a huge range of *compounds*. In nature, atoms tend to bond together and form a molecule whenever the resulting compound is more stable than the individual elements.

The number of electrons in an atom's outer shell are called *valence electrons*. They strongly influence its tendency to combine with other atoms. Based on experience, chemists have assigned to every element in the periodic table one or more numbers indicating its ability to react with other elements. These numbers are determined by the number of valence electrons, so they are called the element's *valences*.

As elements combine to form compounds, valence electrons are transferred from the outer shell of one atom to that of another, or they are shared by the outer shells of the combining atoms. This rearrangement of electrons produces chemical bonds. When electrons are transferred, the process is called *ionic bonding*. If electrons are shared, it is called *covalent bonding*.

Valence indicates the actual number of electrons that an atom gains, loses, or shares in bonding with one or more other atoms. For example, if an atom loses one electron in a reaction forming an ionic bond, then it has a valence of +1; if an atom must receive one electron to form an ionic bond, then it is said to have a valence of –1. This number varies for different atoms depending on several factors, such as the conditions under which the reaction occurs and the other elements involved. Electrons are not lost or gained while forming a covalent bond, so the valence of an atom involved in this type of bond is expressed without a plus or minus sign. Table 5-1 lists common valences for a number of elements.

Chemical Formulas and Notation

The simplest molecules combine only one type of atom, as when two atoms of oxygen combine to form O_2. The subscript indicates the number of atoms of a

Element	Common Valences	Element	Common Valences
Aluminum (Al)	+3	Lead (Pb)	+2, +4
Arsenic (As)	+3, +5	Magnesium (Mg)	+2
Barium (Ba)	+2	Manganese (Mn)	+2, +4
Boron (B)	+3	Mercury (Hg)	+1, +2
Bromine (Br)	−1	Nitrogen (N)	+3, −3, +5
Cadmium (Cd)	+2	Oxygen (O)	−2
Calcium (Ca)	+2	Phosphorus (P)	−3
Carbon (C)	+4, −4	Potassium (K)	+1
Chlorine (Cl)	−1	Radium (Ra)	+2
Copper (Cu)	+1, +2	Selenium (Se)	−2, +4
Chromium (Cr)	+3	Silicon (Si)	+4
Fluorine (F)	−1	Silver (Ag)	+1
Hydrogen (H)	+1	Sodium (Na)	+1
Iodine (I)	−1	Strontium (Sr)	+2
Iron (Fe)	+2, +3	Sulfur (S)	−2, +4, +6

Table 5-1 Valences of Various Elements

particular element. Similarly, two atoms of hydrogen combine with one atom of oxygen to form the compound H_2O. The notation "H_2O" is called the chemical *formula* for water.

Formulas are combined with certain symbols to form chemical *equations* that describe chemical reactions. When the solid sodium chloride (table salt) dissolves in water, the compound breaks apart into ions and the formula for this reaction is

$$NaCl + H_2O \rightarrow Na^+ + Cl^- + H^+ + OH^-$$

This type of an equation always balances. That is, the left and right sides contain an equal number of like atoms. This reflects the principle that matter is neither created nor destroyed, so the number of atoms of each element going into the reaction must equal the number coming out.

Often, chemical equations are simplified by showing the complete compound formulas on each side instead of showing specific ions:

$$Ca(HCO_3)_2 + Ca(OH)_2 \rightarrow 2CaCO_3 + 2H_2O$$

This equation introduces some new notation and a new concept. The parentheses enclose chemical formulas for two *radicals*. A radical is a group of atoms bonded together into a unit that reacts like a single atom with other atoms. The subscript numbers outside the parentheses indicate the number of radicals involved in the reaction.

Also, the number 2 in front of $CaCO_3$ and H_2O, called a *coefficient*, indicates the number of molecules of each compound involved in the reaction. If no coefficient is shown, only one molecule of the compound is involved. Therefore, the equation above indicates that one molecule of calcium bicarbonate reacts with one molecule of calcium hydroxide to form two molecules of calcium carbonate and two molecules of water.

This information can be used to add the elements on each side of the equation to confirm that the equation is balanced. On the left, the first compound, calcium bicarbonate ($Ca(HCO_3)_2$) is made up of one atom of calcium, two of hydrogen, two of carbon, and six of oxygen. Similarly, calcium hydroxide (or lime, $Ca(OH)_2$) includes one atom of calcium and two hydroxyl ions (OH). Each hydroxyl ion includes one atom each of oxygen and hydrogen, for a total of one atom of calcium and two atoms each of oxygen and hydrogen. The grand total on the left side of the equation is two calcium atoms, eight oxygen, four hydrogen, and two carbon.

Carbon atoms have a special ability to form covalent bonds with other carbon atoms and with many other elements. This can create long chain molecules leading to the production of the myriad compounds that are the substance of life.

Acids and Bases

Some substances release hydrogen ions (H^+) when they are mixed with water. This creates an acid, and the reaction produces an acidic solution. For example, when sulfuric acid (H_2SO_4) is mixed with water, many of the molecules dissociate (come apart), forming H^+ and SO_4^- ions. This release affects the characteristics of the solution, giving it a tendency to oxidize.

Other substances produce hydroxyl ions (OH^-) when they dissociate in water. These substances are called *bases*, and the reaction produces an alkaline solution. Examples that are familiar to water treatment personnel are lime ($Ca(OH)_2$), caustic soda (sodium hydroxide, $NaOH$), and household ammonia (NH_4OH).

The acidic or alkaline character of a solution is measured on the pH scale, shown in Figure 5-2. Measurement at the low end of the pH scale represents an increasingly acidic solution; measurement at the high end represents an increasingly alkaline solution. A pH value of 7 indicates a neutral solution, neither acidic nor alkaline. That is the level of pure water.

Acids and bases neutralize one another, so pH can be adjusted upward by adding a base or downward by adding an acid. This capability is important for the

High Concentration of H^+ Ions	H^+ and OH^- Ions in Balance	High Concentration of OH^- Ions
0—1—2—3—4—	5—6—7—8—9—	10—11—12—13—14
Pure Acid	Neutral	Pure Base

Figure 5-2 Acids and Bases on the pH Scale

effectiveness of many water treatment processes. For example, a water's pH rises substantially during softening, so it must be reduced again by adding an acidic substance.

Organic Chemistry

Compounds that are produced by life processes are called *organic*. Over the last century, chemists have synthesized other compounds that are similar to the molecular structure of carbon atoms and are considered to be organic in nature. This list includes plastics, petroleum products, drugs, and textiles.

Many elements and compounds attach to the carbon atom backbones of organic molecules. The vast diversity of these structures has produced a special nomenclature. The prefix in a substance name indicates the number of carbon atoms in the molecule's chain. A name beginning with *meth-* refers to a chemical with one carbon atom. Other prefixes are *eth-* (two carbons), *prop-* (three), *but-* (four), *pent-* (five), *hex-* (six), *hept-* (seven), *oct-* (eight), *non-* (nine), *dec-* (ten), and so forth. In addition, names of some organic chemicals include suffixes that reveal the number of bonds between carbon atoms: *-ane* (one bond), *-ene* (two), *-yne* (three).

The names of some major chemical groups denote similar compounds. Often these compounds attach to longer organic carbon molecules. The list includes: alkanes, alkenes, alkynes, aldehydes, ketones, amines, amides, alcohols.

Suffixes sometimes denote the presence of compounds in these groups. The suffix *-ol* denotes the presence of a compound from the alcohol group, each of which includes one hydroxyl radical (chemical formula: OH). Other suffixes include *-one* for ketones, *-al* for aldehydes, *-amine* for amines, *-amide* for amides, etc. Figure 5-3 illustrates the structures and names of some organic compounds.

The Nitrogen Cycle

Many organic chemicals follow cycles throughout nature. For example, they are assimilated by microorganisms, plants, and animals through metabolic processes and then are returned to the soil, water, or air. One of the most important metabolic cycles is the nitrogen cycle, which is dependent on microbial activity, shown in Figure 5-4. Nitrogen compounds support plant growth and, therefore, animal life. They are preserved in the environment through living plant and animal matter to

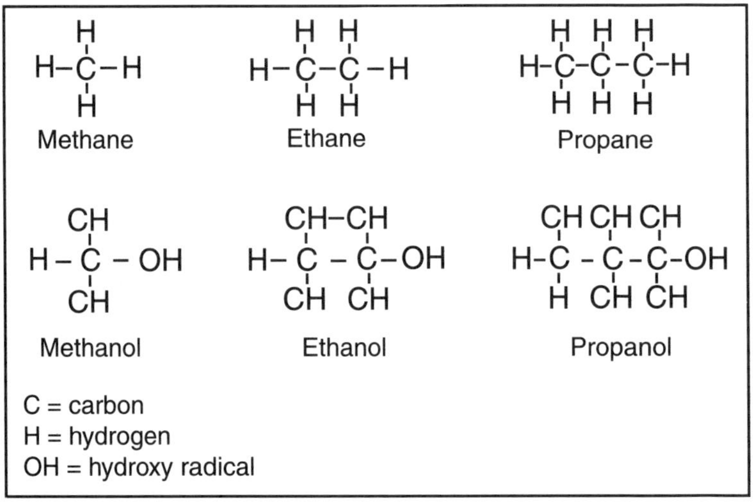

Figure 5-3 Structures and Names of Sample Organic Molecules

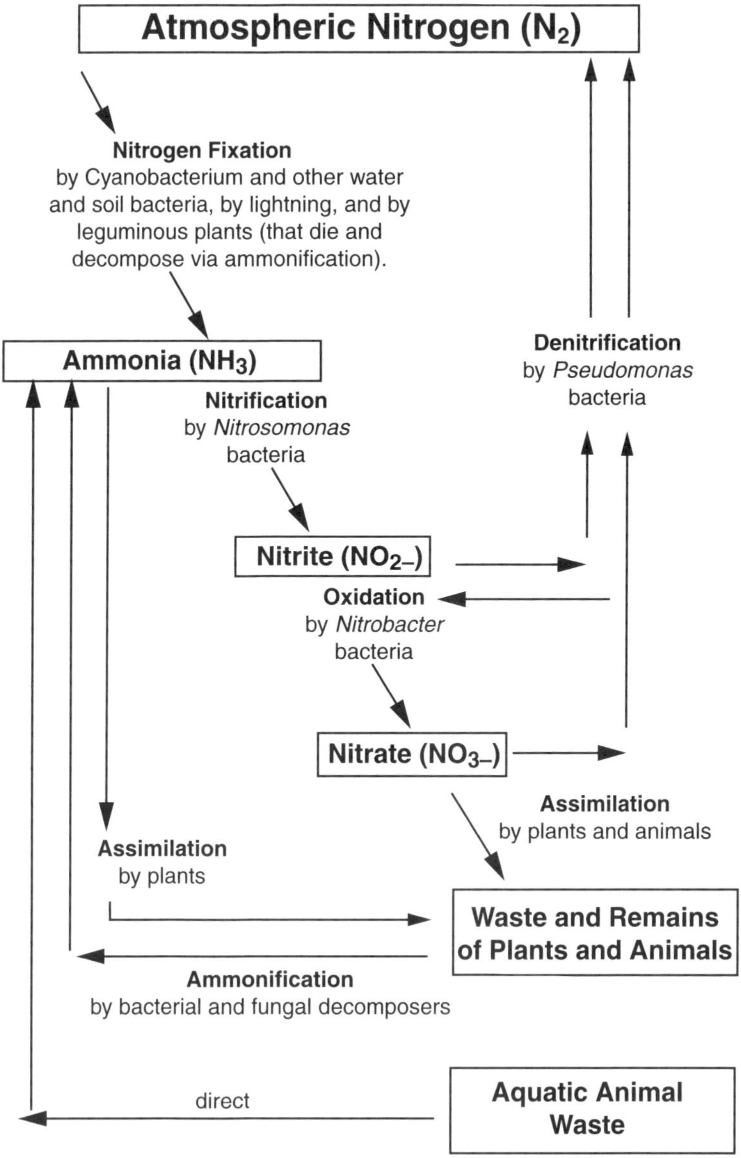
Atmospheric Nitrogen (N_2)
Nitrogen Fixation
by Cyanobacterium and other water and soil bacteria, by lightning, and by leguminous plants (that die and decompose via ammonification).
Denitrification
by *Pseudomonas* bacteria
Ammonia (NH_3)
Nitrification
by *Nitrosomonas* bacteria
Nitrite (NO_{2-})
Oxidation
by *Nitrobacter* bacteria
Nitrate (NO_{3-})
Assimilation
by plants and animals
Assimilation
by plants
Waste and Remains of Plants and Animals
Ammonification
by bacterial and fungal decomposers
direct
Aquatic Animal Waste

Figure 5-4 Nitrogen Cycle

dead organic matter in various stages of decomposition, once again nurturing plant growth. During this process, some strains of bacteria found in soils convert naturally available ammonia (NH_3) to nitrite (NO_2^-) and then to nitrate (NO_3^-).

Because nitrogen is important to plant growth, many commercial fertilizers contain sodium nitrate ($NaNO_3$) and potassium nitrate (KNO_3). Storm runoff carries both naturally produced and artificially introduced nitrates and nitrites into rivers and lakes used as public water supplies. Both compounds have implications for public health, and they are regulated by the US Environmental Protection Agency. For example, consumption of nitrite compounds by infants may induce methemaglobinuria (blue-baby syndrome), which dangerously impairs the oxygen-carrying hemoglobin of their blood. Breast-fed babies do not have enough acid in their stomachs to inhibit the growth of bacteria that reduce nitrate to nitrite. Once babies begin eating solid food, they develop levels of stomach acid that inhibit bacteria growth, and they lose their susceptibility to the disease.

In addition, excessive nitrate and nitrite levels may cause large-scale environmental problems. The Mississippi River has high levels of nitrates. During spring runoff, the influx of these nutrients into the Gulf of Mexico causes the growth of great blooms of algae, bacteria, and other microorganisms. The metabolic processes of these organisms deplete the oxygen in the water causing hypoxia, which kills fish and other aquatic animals and significantly impacts the fishing and tourism industries.

To maintain compliance with regulations and to ensure that they supply healthy products, water treatment plants must monitor the levels of nitrate and nitrite compounds in their finished water.

Microbiology in Water Treatment

Removing or inactivating microorganisms is an important goal of almost any water treatment process. The addition of lime ($Ca(OH)_2$) removes suspended solids and dissolved chemicals that cause hardness. This process of softening the water also helps kill or inactivate pathogenic microorganisms. Chlorine or other disinfectants are added to filtered water specifically to prevent the spread of pathogens. Some organisms, notably the parasites *Giardia lamblia* and *Cryptosporidium parvum*, are resistant to disinfection, but effective coagulation and filtration can help to remove them, as can membrane filtration.

Softening

Water treatment often involves application of chemical principles to remove calcium bicarbonate, the cause of *carbonate hardness*. This is accomplished by adding slaked lime to the water:

$$\underset{\text{calcium oxide (lime)}}{CaO} + H_2O \rightarrow \underset{\text{calcium hydroxide (lime slurry or "slaked lime")}}{\mathbf{Ca(OH)_2}} + \text{Heat}$$

$$\mathbf{Ca(OH)_2} + \underset{\text{calcium bicarbonate (carbonate hardness)}}{Ca(HCO_3)_2} \rightarrow \underset{\text{calcium carbonate (lime sludge)}}{\mathbf{2CaCO_3}\downarrow} + 2H_2O$$

$$\mathbf{Ca(OH)_2} + \underset{\text{(another component of hardness)}}{Mg(HCO_3)_2} \rightarrow \mathbf{CaCO_3} + MgCO_3\downarrow + 2H_2O$$

$$\mathbf{Ca(OH)_2} + MgCO_3 \rightarrow \mathbf{CaCO_3}\downarrow + \mathbf{Mg(OH)_2}$$

Note that the downward-pointing arrows (↓) indicate that the compounds are insoluble and precipitate out of the water.

Addition of slaked lime also aids in the removal of noncarbonate hardness:

$$\mathbf{Ca(OH)_2} + \underset{\substack{\text{magnesium sulfate} \\ \text{(noncarbonate hardness)}}}{MgSO_4} \rightarrow \mathbf{Mg(OH)_2} \downarrow + CaSO_4$$

$$\underset{\substack{\text{sodium carbonate} \\ \text{(soda ash)}}}{\mathbf{NaCO_3}} + CaSO_4 \rightarrow \mathbf{CaCO_3} \downarrow + NaSO_4$$

These chemical reactions raise the water's pH to between 10 and 11. The vast majority of viruses, bacteria, protozoa, and multicellular organisms cannot survive in this condition, so water softening is a powerful disinfectant. Carbon dioxide (CO_2) is added after softening to reduce the water's pH.

Disinfection

Not all water treatment plants use lime softening, and those that do also introduce a variety of disinfectants to ensure that active pathogens are not distributed in finished water. Removal of microorganisms actually begins with coagulation and flocculation. As these processes create flocs of suspended solids, some microorganisms are trapped with those particles and removed when the solids settle out during clarification.

More are removed when clarified water passes through granular filters. In rapid sand filters, the first few inches of the granular media trap remaining flocs, algae, and other materials, which in turn capture some microorganisms. The schmutzdecke performs the same function in a slow sand filter.

This physical removal process is very efficient in purifying the water, especially if treatment includes softening. In fact, it may be the most effective way within a conventional treatment plant to remove some pathogens that are resistant to disinfection chemicals. Following filtration, the final stage of treatment usually includes disinfection using chlorine, chlorine dioxide, chloramine, or ozone. Some water treatment plants employ new disinfection processes using ultraviolet light (UV) or membrane filtration. Choices among disinfectant practices depend on several factors such as cost, safety, convenience, effectiveness, and the condition of the water being treated.

Chlorine (Cl_2) is the most widely used disinfectant in water treatment. It is relatively cheap, easy to use, and effective at killing most microorganisms present in the water. In addition, some of the chemical remains in the finished water as a free-chlorine residual that prevents bacterial regrowth in the distribution system.

When chlorine is added to pure water, it reacts as follows:

Cl_2	+	H_2O	→	HOCl	+	HCl
chlorine		water		hypochlorous acid		hydrochloric acid

The chlorine combines with the water to produce hypochlorous acid (HOCl), a weak acid that easily penetrates into and kills bacteria. This action makes chlorine an effective disinfectant. HOCl is also one of two chlorine compounds that acts as free available chlorine residual. However, some of the HOCl dissociates as follows:

HOCl	→	H^+	$+OCl^-$
hypochlorous acid		hydrogen	hypochlorite ion

The hydrogen produced in this reaction neutralizes alkalinity and lowers pH, while the hypochlorite ion (OCl^-) is a second type of free available chlorine residual. Its disinfectant action is not as effective as that of HOCl, but it does help to kill microorganisms.

A number of conditions influence the effectiveness of disinfection using chlorine, including pH, water temperature, and contact time. The water's pH when Cl_2 is introduced strongly influences the ratio of HOCl to OCl^-. Low pH values favor formation of HOCl, the more effective free residual, while high pH favors formation of OCl^-. Similarly, low water temperature slightly favors formation of HOCl.

Contact time is important, along with the concentration of chlorine in the water, in determining the effectiveness of these compounds. Either a longer time or a rising concentration increase disinfectant effect.

The chemical reactions, which have been described, occur when chlorine is added to pure water. Water being treated with disinfectant also contains other substances, however. For example, Cl_2 reacts with ammonia present in the water to form *chloramine* compounds:

NH_3	+	$HOCl$	→	NH_2Cl	+	H_2O
ammonia		hypochlorous acid		monochloramine		

NH_2Cl	+	$HOCl$	→	$NHCl_2$	+	H_2O
monochloramine		hypochlorous acid		dichloramine		

$NHCl_2$	+	$HOCl$	→	NCl_3	+	H_2O
dichloramine		hypochlorous acid		trichloramine		

The water's pH and the amount of ammonia present determine whether one chloramine compound or more than one are formed.

Monochloramine and dichloramine act as disinfectants, although they are not as effective as free available chlorine, such as HOCl. If contact time is sufficient, chloramines can do an acceptable job of disinfection. However, dichloramine and trichloramine compounds may produce taste-and-odor problems.

Disinfection with chlorine requires additional planning for water with high levels of natural organic material (NOM). Chlorine reacts with NOM to form undesirable disinfection by-products (DBPs). Some of these compounds, especially trihalomethanes (THMs), are regulated as potential carcinogens by USEPA under the Disinfectants/Disinfection By-products Rule. To avoid creating these compounds, a utility must minimize residual organic material in the water before disinfection and carefully control chlorine usage. Effective coagulation, flocculation, sedimentation, and filtration remove many organic compounds, so chlorine will not react with the compounds to generate THMs. Chloramine reactions do not produce THMs.

Chlorine dioxide (ClO_2) is another very effective disinfectant, and its reactions produce fewer DBPs than those of chlorine. Chlorine dioxide is also more costly and more hazardous to use, however.

Ozone has become a popular disinfectant, despite its high cost, with utilities that have problems with protozoa. In particular, ozone treatment is one of the few effective methods for inactivating *Cryptosporidium*. After ozonation, however, the water contains no residual disinfectant to protect against regrowth of pathogens in the distribution system. Also, ozone cannot be stored, so it must be generated on-site as needed.

Ultraviolet light is another disinfection technology used by some utilities. Under favorable conditions, it inactivates almost all microorganisms without producing undesirable THMs or other DBPs. Once practical only for small facilities, the technology is now being considered at large water utilities. The primary drawback to this treatment is the potential for the lightbulbs to become coated with light-obscuring material, which prevents the UV light from reaching and killing the organisms. Careful maintenance is needed to ensure efficient operation. Also, turbidity in the water can shield organisms from the UV light, so this technology is practical only in very clear waters. Finally, UV treatment does not leave any disinfectant residual.

Membrane filtration is an effective technology for physical removal of microorganisms from water. After water passes through granular media filters, it is fed through membranes with pores of specific size. Membrane-based water treatment processes (listed in order from larger to smaller pore sizes) include microfiltration, ultrafiltration, nanofiltration, and reverse osmosis. Each removes a progressively finer particle. Ultrafiltration is usually sufficient to remove all microorganisms. No disinfectant residual remains in the water after membrane treatment.

Appendix A

Scientific Nomenclature, Scientific Notation, and Units of Measure

Scientific nomenclature is arranged from the most general to the most specific.

Example: Family *Enterobacteriaceae*
Genus *Escherichia*
Species *coli*
Strain *O157:H7*

Use the genus and species to routinely denote microorganisms. The first letter of the genus is always capitalized, and the species name is always in lower case: *Escherichia coli*. The names should also be italicized.

Group names such as *bacteria* or *enteroviruses* are not considered scientific names and begin with lowercase letters.

Scientific Notation

Scientific notation is a convenient way to express large numbers by writing them in powers of ten:

$1 \times 10^1 = 10$

$1 \times 10^2 = 100$

$1 \times 10^3 = 1{,}000$

$1 \times 10^4 = 10{,}000$

$1 \times 10^5 = 100{,}000$

$1 \times 10^6 = 1{,}000{,}000$

Within this system, for example, 1,200,000 is expressed as 1.2×10^6. The superscript 6 indicates that the decimal point must be moved to the right six places. Similarly, 0.004 is expressed as 4.0×10^{-3}. The superscript –3 indicates that the decimal point must be moved to the left three places.

Units of Measurement

Prefix	Abbreviation	Number
Mega-	M	One million
Kilo-	k	One thousand
Deci-	d	One tenth
Centi-	c	One hundredth
Milli-	m	One thousandth
Micro-	µ	One millionth
Nano-	n	One billionth
Pico-	p	One trillionth

Parts per million (ppm) = milligrams per liter (mg/L) = 1 gram per million milliliters

References

American Public Health Association, American Water Works Association, and Water Environmental Federation. 1998. *Standard Methods for the Examination of Water and Wastewater*, 20th edition. Washington, D.C.: APHA.

American Water Works Association. 1995. Manual M7: *Problem Organisms in Water: Identification and Treatment.* Denver, Colo: AWWA.

______. 1997. Manual M12: *Simplified Procedures for Water Examination*, 2nd edition. Denver, Colo.: AWWA.

______. Manual M48: *Waterborne Pathogens*, 1st edition. 1999. Denver, Colo.: AWWA.

______. Water Supply Operations, Part V: *Basic Science Concepts and Applications*. 1995. Denver, Colo.: AWWA.

Coleman, Gordon J. and David Dewar. 1997. *The Addison-Wesley Science Handbook*. Boston: Addison-Wesley.

Doyle, Michael, Larry R. Beuchat, Thomas J. Montville. 1997. *Food Microbiology Fundamentals and Frontiers.* Washington, D.C.: ASM Press.

Forbes, Betty A., Daniel F. Sahm, Alice S. Weissfeld. 1994. *Bailey and Scott's Diagnostic Microbiology*, 10th edition. Musby, Inc.

Kerri, Kenneth D. 1994. *Water Treatment Plant Operation Volume 1*. Sacramento, California State University: School of Engineering.

Lisle, John. 1994. *An Operator's Guide to Bacteriological Testing*. Denver, Colo.: AWWA.

Murray, Patrick R. 1995. *Manual of Clinical Microbiology.* Washington, D.C.: ASM Press.

Index

Note: *f.* indicates figure; *t.* indicates table.

K

L

M

S

T